MEDIA AND COMMUNICATIONS - TECHNOLOGIES, POLICIES AND CHALLENGES

RADIO-FREQUENCY IDENTIFICATION (RFID)

EMERGING TECHNOLOGIES, APPLICATIONS AND IMPROVEMENT STRATEGIES

MEDIA AND COMMUNICATIONS - TECHNOLOGIES, POLICIES AND CHALLENGES

Additional books in this series can be found on Nova's website under the Series tab.

Additional e-books in this series can be found on Nova's website under the e-book tab.

MEDIA AND COMMUNICATIONS - TECHNOLOGIES, POLICIES AND CHALLENGES

RADIO-FREQUENCY IDENTIFICATION (RFID)

EMERGING TECHNOLOGIES, APPLICATIONS AND IMPROVEMENT STRATEGIES

XINXIN FAN
EDITOR

nova publishers
New York

For permission to use material from this book please contact us:
Telephone 631-231-7269; Fax 631-231-8175
Web Site: http://www.novapublishers.com

Additional color graphics may be available in the e-book version of this book.

LIBRARY OF CONGRESS CATALOGING-IN-PUBLICATION DATA

ISBN: 978-1-63117-750-7

Published by Nova Science Publishers, Inc. † New York

CONTENTS

PREFACE

Radio Frequency Identification (RFID) technology enables the unique, automatic and non-contact identification of objects using radio waves. It is an exciting, rapid growing, multidisciplinary field with emerging technologies and applications across multiple economic sectors. As one of key enablers of ubiquitous computing and Internet of Things (IoTs), RFID systems are being deployed throughout the world at a staggering rate and impacting peoples' life in significant ways. It is estimated that the RFID market will increase from $6.98 billion in 2012 to $7.88 billion in 2013, and reach $23.4 billion in 2020. RFID technology has achieved widespread success in a multitude of domains, ranging from manufacturing, asset tracking, smart home appliances, supply chain management, retailing, payment systems, to security and access control. Standards organizations such as International Organization for Standardization (ISO) and EPC global have made a great effort to specify various applications and ensure the interoperability of RFID systems around the world during the past few years. While RFID technology brings unique features and stimulates innovative applications, there are significant research challenges that need to be addressed before the widespread adoption of RFID systems.

Significant R&D work has been undertaken over recent years on RFID systems in both academia and industry, with emphasis on new hardware design of RFID readers and tags, sophisticated software infrastructures for supporting IoTs, efficient power harvesting technology, lightweight cryptographic algorithms and protocols for protecting security and privacy, sensor embedded RFID technology, leading-edge innovative RFID applications as well as their business aspects. This book consists of 9 chapters and aims to bring together RFID researchers and practitioners from academia and industry to present the state-of-the-art developments, trends, industrial applications and research challenges related to RFID technology. The first 4 chapters focus on the RFID applications and research tools, whereas the last 5 chapters explore the security and privacy issues in RFID systems. This book represents a comprehensive and useful reference for recent progress in RFID technology, applications, security and privacy, which is written for graduate students, researchers, educators, and engineers working in the field of wireless networks and mobile communications.

I would like to express our gratitude to the authors of book chapters for their excellent work, which ensures that this book presents the latest progress in RFID technology in a timely manner. Special thanks also go to Carra Feagaiga of Nova Science Publishers, Inc. for the support, patience and professionalism during the publication process of this book. Last but not least, my acknowledgements also go to Prof. Guang Gong from the Department of Electrical and Computer Engineering, University of Waterloo, Canada, for her constant encouragement and support throughout this book project.

Xinxin Fan, Ph.D.
Department of Electrical
and Computer Engineering
University of Waterloo
Waterloo, Ontario, Canada
x5fan@uwaterloo.ca

In: Radio-Frequency Identification (RFID)
Editor: Xinxin Fan, pp. 1-44
ISBN: 978-1-63117-750-7

Chapter 1

DESIGN AND APPLICATION OF RFID TECHNOLOGY IN THE CONTAINER TRANSPORT SYSTEM

***Jifeng Qian**, Zhandong Liu and Xia Xie**
Department of Military Logistics, Tianjin Transportation University, China

Abstract

The use of radio frequency identification (RFID) can improve the efficiency of container multimodal transportation. How to design the RFID tag, reader-writer and middleware system, how to select the frequency and type of RFID system used in container multimodal transportation, where to install the RFID tags and readers, all these mentioned above are discussed in this chapter. At last the function and application of RFID system in container transportation are analyzed, and the evaluation of the RFID label system is done. The result suggests that the container RFID system can promote the realization of container transport organization towards seamless cohesion and RFID technology will be widely used in the container transportation in the future.

Keywords: RFID Technology; RFID Reader; RFID Tag; Container Transportation

* E-mail address: qianjifeng@sina.com; Address: No.1 Dongjuzi, Hedong District, Tianjin, China (Corresponding author)

Introduction

RFID is the abbreviation of "Radio Frequency Identification" and it is a non-contact automatic identification technology. During World War II, the British aircraft were equipped with high power consumption of active tags to receive radar signals so to distinguish between enemy and friendly forces, that is the time RFID was first used. In the late 1970s, the U.S. government transferred RFID technology to the civil use through the Los Alamos Scientific Laboratory. Karen Coyle studied the need, the role, as well as problems of RFID for the Library application, Harry K.H Chow, King Lun Choy and W. B. Lee etc. designed a warehouse resource management system based on RFID technology. Nowadays, RFID technology has been widely used in the field of industrial automation, transportation control and management, immigration passport safety control, livestock management, and supply chain management.

More applications of RFID in Container transportation were used in the United States and Europe. On the military side, in order to prevent shipping lines, the chain of international trade breaking from terrorist attacks after the 911 affair, the Customs Trade joint anti-terrorism plan and Container Security Agreement were carried out by the United States government to strengthen the management of cargo security in port. The former aims at prevention of terrorism customer relationships and formulates the container seal standard ISO17712.The latter aims to strengthen the inspection of containers of hazardous material smuggling. The U.S. government designated about 58 ports must be added to this agreement in the field of global trade. The Chinese ports of Shanghai, Hong Kong, Shenzhen and others have also incorporated. At the military aspects, the United States Munitions Command applies EPC Gen2 tags to track containers. In the aspect of maritime logistics, the South Carolina State Ports Authority applied RFID technology to track the management of containers in March 2007 so that shippers can track input and output containers at the port. The California container manufacturer teamed with the customer to test the passive RFID tag. It was firstly applied in liquid loading boxes and drums and expected to gradually extend the production line in order to achieve tracking management in the packaging process and supply chain of the container. The third-party logistics provider Jobstl of Austria applied RFID technology to track containers, so the container tracking technology has been greatly improved, thereby solving the problem of container loss.

The Korean government launched the "smart container" project at the end of 2004 and it is also known as RFID maritime logistics named under the monitoring of transport safety system software. The background personnel of management can get all kinds of information in real-time, so to improve the international trade efficiency and security between South Korea and other countries, the Busan port of South Korean used the SaviTrak service based RFID of Savi Networks Company for container and cargo security and management in January 2007.

Components, Working Principle and Workflow of RFID System

Components of RFID System

RFID technology is a non-contact automatic identification technology, there are four components consisted in the most complete RFID system, such as tags, readers, antennas and computer (database) system, the composition of the system model as shown in Figure 1.

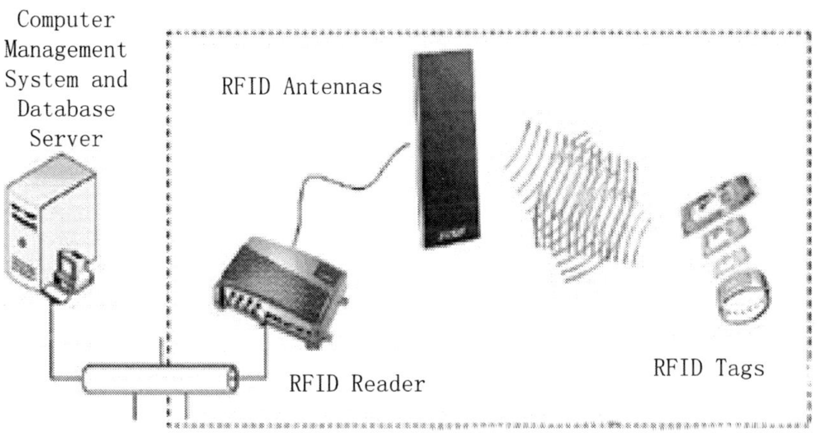

Figure 1. Constitute of RFID System.

RFID tags are constituted of the coupling element (antenna) and chip, each tag has a unique electronic code to identify the target object when attached to the object. The reader can read or write the information on the label, so it can be designed for handheld or fixed. RFID antenna is used for transmitting RF signal between tags and readers.

(1) Tags

The RFID tag is an electronic apparatus in which the identification data can be stored in the radio frequency identification system and it is usually installed in the object to be identified. When the reader queries, it will send data to a non-contact manner for the reader to read, perform tasks including reading memory, writing memory, or deal with memory data, such as encryption. It has the following features:

1) Stored functions, the relevant information about the object can be stored in to be identified.

2) Data stored in the electronic tag can be read or written in a certain working environment and conditions.

3) Maintain the integrity of identification and related information of the identifying object.

4) Transmitting timely to the reader after data encoding.

5) Programmable, and persistent data cannot be modified after programming.

6) Having a determined period of using, without maintenance during the period of using.

7) For an active tag, the reader can display the working status of the battery.

(2) Reader

The reader is the device for the capturing and processing of the RFID tag data, collecting information through the RF signal to complete the data transfer and it is the main member of the system constituting. In the process of the system working, the reader usually transmits RF energy to form an electromagnetic field within a region, the operating distance varies depending on the size of the transmit power. It is triggered when the tags through this region and send the data stored in the label or rewrite the data stored in the label in accordance with the instruction of the reader. The reader can receive the data sent by the tags or write data to the tags and communicate with the computer network through a standard interface. Specifically, the readers have the following functions.

1) Communication functions between the readers and tags. The reader and the RF tag can communicate at the technical conditions prescribed.

2) Communication between the reader and the computer through a standard interface such as RS232. The reader can be connected to the computer network via a standard interface and provide relevant information (including the ID of the

reader, the time of reader reading tags and tag information read by the reader), so to achieve the operation of the multi-reader in the system network.

3) Readers can read and write multi-tag in the reading region at the same time and has anti-collision function.

4) Readers are applicable for fixed and mobile tag reading.

5) Readers can check the error message in the reading and writing process.

6) For active tags, the reader is able to identify the battery information, such as electricity.

(3) RFID Antenna

The antenna is an apparatus which can convert electromagnetic wave received to a current signal or convert the current signal to the electromagnetic wave. The reader transmits energy to form the electromagnetic field through an antenna in the radio frequency identification system so to identify the radio-frequency tag by the electromagnetic field. Thus, the antenna communicates transmission channel of electromagnetic wave between the radio-frequency tag and the reader. The range of the electromagnetic field formed by the antenna is the readable area of the RFID system. Any RFID system transmits and receives RF signals by at least one antenna. In the UHF and microwave systems, the planar antenna, Omni directional panel antenna, horizontal planar antennas and vertical panel antenna are widely used.

(4) Computer Database Systems

Computer database systems mainly complete the data storage, management and reading and writing control of the radio frequency tags, and provide the data obtained by the reader to the application. In addition to the application system itself, the RFID application framework based middleware typically between the reader and the system is more important.

A complete RFID application system also includes the RFID middleware, the corresponding anti-collision mechanism and the data processing mechanism, as is shown in Figure 2.

Working Principle of RFID System

The radio frequency identification technology is developed based on communication technology. The basic principle is using the radio frequency signal and its spatial coupling (inductive or electromagnetic coupling) and transmission characteristics to identify information of the object to be identified automatically. The electromagnetic coupling between the RF tag and the reader is divided into two cases, close-inductively coupled with an electromagnetic coupling and the distant backscattered electromagnetic coupling.

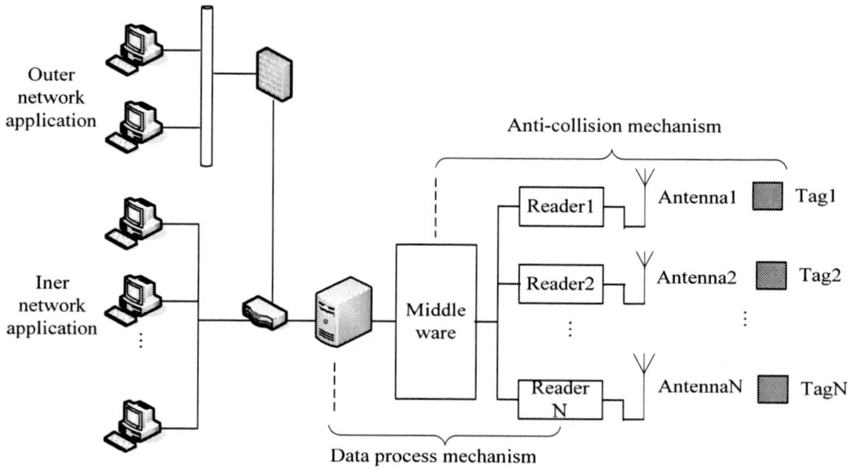

Figure 2. Complete Constituter of RFID System.

(1) Inductively Coupled

In the inductive coupling, the coupling medium is the spatial magnetic field. The antenna of the reader corresponds to a primary winding of the transformer and the antenna of the radio-frequency tag is equivalent to the secondary winding of the transformer. The coupling magnetic field forms a closed loop between the secondary windings and the primary winding of the radio-frequency tag reader. The part of the magnetic force lines of the emission magnetic field pass through the antenna coil of the reader and antenna coil of tag some distance away from the antenna coil of the reader. By electromagnetic induction, a voltage is generated in the antenna coil of the tag rectifying the voltage as the power of tags. At the same time, the capacitor and the antenna coil of the reader parallel resonant frequency capacitance and inductance of the antenna coil with the reader together formed transmit frequency to match the parallel resonance frequency. While tune resonant frequency formed by the oscillation circuit of the transponder antenna coil and the capacitor to the emission frequency of the reader, a transmission mode of communication can be formed. It is generally suitable for low-frequency short-range radio frequency identification system and the typical operating frequency are 125kHz, 225kHz and 13.56 MHz. Actually the role of the distance is less than 1m, the typical distance is from 10 to 20cm. Accordingly, the inductive coupling mode is the transformer model of low frequency close exposure to radio frequency identification system. Its principle is shown in Figure 3.

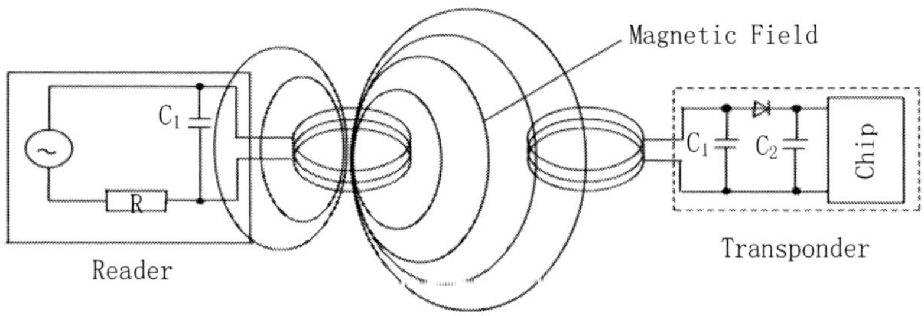

Figure 3. Operating Principle of the Inductively Coupled Transponder.

(2) Electromagnetic Backscatter Coupling

In the electromagnetic coupling method, RF energy is generated by the reader and sent to the orientation of the space through the antenna of the reader as the mode of electromagnetic waves so as to form the effective recognition region for the reader. The radio-frequency tag extracts energy from the electromagnetic fields emitted by the reader antenna, and the data stored in the radio-frequency tag is transmitted to the reader by the internal circuit of the tag and the tag antenna as the backscattering mode. The specific process is described as follows, the power of the electromagnetic wave P_1 emitted by the reader via the antenna after attenuate in free space whereby the part of P_1'reaches the antenna of the electronic tag as the high-frequency power at the antenna interface for electronic tags. At the same time, P_2' (the part of the power P_1') is reflected by the antenna of tags after attenuate in free space, whereby P_2 (the part of P_2') is received by the reader antenna. In the course of this work, the turning on and off of the additional resistor R_L paralleled with the tag antenna will affect the data transmission between the electronic tag and the reader. Modulation of the amplitude of the reflected power P_2 of the electronic tag can be achieved by adjusting the resistor R_L to achieve the correct data transmission. It is generally applicable to the work of the high-frequency, microwave remote, radio frequency identification system. The typical operating frequencies are 433MHZ, 915MHz, 2.45GHz and 5.8GHz. Therefore, the electromagnetic backscatter coupling mode is UHF or microwave radar mode of long-distance non-contact radio frequency identification system. Its principle is shown in Figure 4.

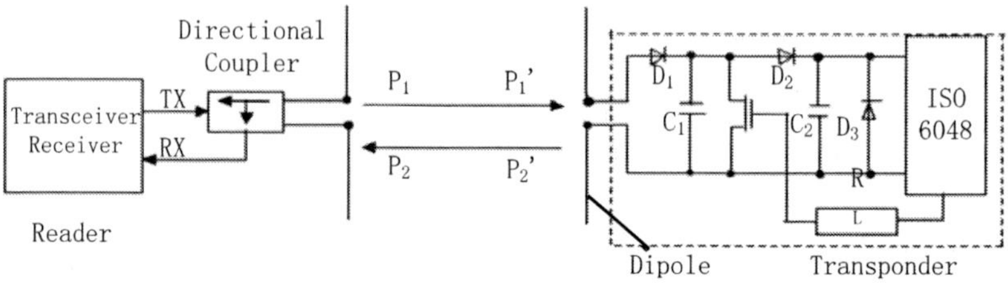

Figure 4. Operating Principle of the Backscatter Transponder.

(3) Comparison between the Two Modes

The difference of inductively coupled and electromagnetic coupling is as follows, while in an inductively coupled way the RF energy is trapped in the inductance coil of the reader around. Through near line alternating closed coil magnetic field communication reader winding with radio frequency tags windings between the RF channel, do not radiate electromagnetic energy into space, and only support the low-frequency communication at close range. While in electromagnetic coupling way, the reader sends RF energy in the form of electromagnetic waves and tags to extract energy from being excited after the transmission of information to the reader. Therefore, supports the long-range UHF or microwave communications.

RFID System Workflow

Readers transmit the RF signal of a certain frequency through receiving antenna of RFID system when the target objects attached with the RFID tag enter into the magnetic field sent by antenna and are identified. The sensed current will be generated in the built-in antenna of tags. The RFID tags, obtain energy by virtue of the induced current and are activated then send the coding information of the product stored in the chip in the system by their built-in antenna and the system's receiving antenna. The system receiver with antenna receives the carrier signal sent from the radio-frequency tag and transmits to the reader via the antenna regulator. The reader demodulates and decodes the received signal then send to the computer data management system performs the correlation process. Computerized data management systems determine the legitimacy of the radio frequency tags based on the logic operation, make the processing and control for different settings, issue the command signal and control the actions of the implementing agencies. The RFID system workflow is shown in Figure 5.

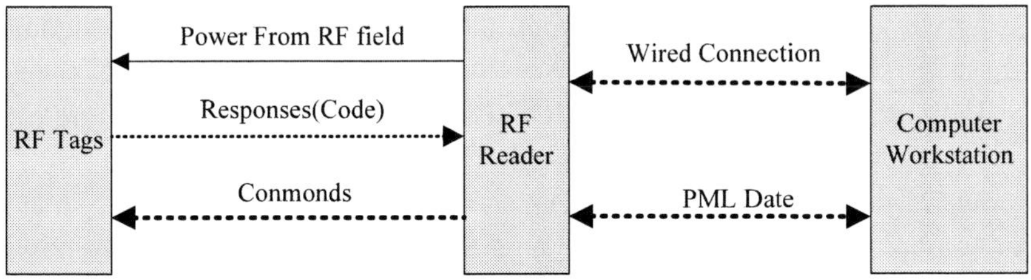

Figure 5. Operating Process of RFID System.

Compared with the application of automatic identification technology in the field of industrial automation, transportation control management, RFID technology application in the container transport has strong particularity.

Firstly, the working conditions are very poor. Electronic labels with the container flow in the sea, terminals, yards and other places. The working environment is harsh, such as wider range of temperature and high temperature over 80℃, air humidity, severe acid corrosion, shock and vibration. Secondly, recognition requirement is higher. Container moves faster and a higher demand for electronic tag identification parameters is put forward, such as quick recognition speed (moving speed>100km/h), long distance (>6m), high accuracy (>99%), large data capacity (>8Kbits) etc. However, the RFID systems are numerous by the larger classification of the indicators and different types of indicators determine the different functions. Then, what kind of RFID system suitable for the application in the container port? Therefore, the classification of RFID systems should be introduced first and then the container port RFID system indicators selected and recommended can be accorded.

Classification of RFID System

RFID systems can be classified according to the characteristics of the radio frequency identification system, the commonly used classification methods are described below:

(1) Label Energy Supply and Data Modulation Classification

1) According to the power supply in the form of the tags, RFID systems can be divided into three categories, active, passive and semi-active systems.

Active tag has a battery installed on it, provide work for all or part energy rely on their own battery. Active tags generally have long reading distance, but they have the shortage that the battery life is limited (usually 3-10 years).

Passive tags have no battery, when they receive the microwave signal sent by the reader, they can convert microwave energy into direct current for their own work. Compared with active systems, passive systems are somewhat more limited in reading distance and adapt to the speed of the moving objects.

Tags of the semi-active system have a battery, but the battery only plays the role of activating the system, after the tags are activated without battery-powered, then can enter the operating mode of the passive tags.

2) According to the data modulation of Label, RFID system can be divided into active, passive and semi-active.

Active system label uses their own energy to take the initiative to send data to the reader; modulation mode can be AM, FM or phase modulation.

Passive system labels have no battery within it, the communication energy of the label must be gotten from the electromagnetic waves emitted by the reader. In case there is an obstacle, passive tags transmit data using the modulation scattering mode, the energy of reader is required to pass through the obstructions twice, whereas in active tag mode, the tag signals transmit through the obstacle only once, and the transmission distance is far.

There is a battery in a semi-active system, but the battery only plays a role of activating the internal digital circuit.

(2) Classification of Operating Frequency

From the actual situation, the operating frequency of the RFID system is also the operating frequency of the tag. The operating frequency of the label determines the system works, communications mechanism, identify distance but also determines the ease of design implementation and cost of the tag and the reader. The basic distribution of the operating frequency of the RF tag in the international arena is shown in Table1.

Table 1. Basic Distribution of the Operating Frequency of the RF Tag

Classification	Range	Common operating frequencies
Low frequency	135 KHz	125 KHz and 134 KHz
High frequency	3 MHz to 30 MHz	13.56 MHz
UHF	300 MHz to 930 MHz	433 MHz and 915 MHz
Microwave	2.45 GHz and above	2.45 GHz and 5.8 GHz

These frequencies all have the corresponding international standard, used in the application in the world recognition. Corresponding to the low frequency are the international standards ISO11784/11785 (radio frequency identification of animals: the code structure / technical concept), ISO14233-1/ISO14233-2 (radio frequency identification of animals: the reader protocol defines / air interface parameters), ISO18002-2 (125 KHz/ 134KHz frequency air interface parameters). Corresponding to the High frequency are the international standards ISO14443/ISO15693 (Near coupling / hydrophobic coupling IC card standard), ISO18003-3 (13.56 MHz frequency air interface parameters). Corresponding to Ultra high frequency are the international standard ISO18006 / ISO18006-7 (860 to 930MHz/433MHz frequency air interface parameters); Corresponding to the microwave band are relevant international standards ISO18006-4 / ISO18006-5 (2.45GHz/5.8GHzfrequency air interface parameters).

Different frequencies of RFID system technical parameters performance comparison are shown in Table 2.

Compared with low frequency systems, the high-frequency systems can identify distance farther, distinguish speed faster and are more suitable for application in occasions of fast-moving.

(3) Accord Timing Sequence of Communication between the Tag and Reader

Timing sequence refers to the working order of the reader and the tag, and divided into two modes. One is RTF (Reader Talks First), means the label sends data to the reader only received special commands. The other is TTF (Tag Talks First) and means the label takes the initiative to send data and the reader records or interrogates further information based on the information sent by the tag in order to form a complete dialogue. Generally the active tags are TTF modes and compared with RTF, TTF mode has the advantages of an agreement with a recognition speed and good anti-collision performance. So it is more suitable for high-speed moving objects occasions.

(4) Classified by Tag Readability

According to the type of memory used within the radio frequency tags, tags can be divided into read-only and read-write types. The read-only label can only be read and not be written in the recognition process, but the contents of the Read-write label can be read out and written by the reader in the recognition process.

Table 2. Technical Parameters Performance Comparison of Different Frequency RFID

Frequency	Low-frequency	Frequency	UHF	Microwave
Typical band	125KHz/134KHz	13.56MHz	433MHz/915MHz	2.45GHz/5.8GHz
Countries and regions	All	Most	Most	Most
Communicate mechanism	Inductively Coupled	Inductively Coupled	Electromagnetic coupling	Electromagnetic coupling
Data transfer rate	Low (8bit/s)	High (64bit/s)	High (64bit/s)	High (64bit/s)
Wavelength	2400m	20m	69cm /33cm	12cm/5cm
Recognition speed	Low (<1m/s)	Middle (<5m/s)	Higher (<50m/s)	High (<60m/s)
Identification distance	under 10cm	10cm~1m	4~6m	25~50cm (Passive) Over 10m (Active)
Label storage capacity	48B~736B		32KB~64KB	
Humid environments	No effect	No effect	Greater impact	General impact
Propagation performance	May penetrate the conductor	May penetrate the conductor	Linear propagation	Linear propagation
Anti-collision performance	Limited	Good	Good	Good
antenna size	Large	Larger	Small	Smaller
Main applications	Access control, animal identification, anti-theft lock library, shelves	Library shelves	Truck, trailer tracking	Toll stations, storage and logistics

Design of RFID System in Container Terminal

Label Choice

For container port, electronic tags include motor vehicle tags and container tags. Both of vehicle and containers can achieve the radio frequency identification using the same technology. Active tags have the advantages of long distance identification, large information storage, wireless reading, writing and programming. Trailers and containers move faster in the port transport, container documents need timely change and supplement, the illegal operations information of the trailers also need on-site processing, so active electronic tags become the preferred application labels. At the same time, active tags can provide power to the sensor in the container transportation, which is in accord with container tracking, while the passive tags are unable to meet. In order to ensure the safety of container transport between ports at home and abroad and monitor the cargo in container, container labels should also be able to record the time, place and temperature when opening the door automatically. The container labels should have the GPS location coordinates and the sensor unit. Taking into account the existing RFID system may reflect the media signal while collecting the container and trailers information by consider using metal made tags for containers and ceramic package tags for trailers.

Reader's Choice

According to the applications of the reader, the reader can be divided into the fixed reader, OEM (Original Equipment Manufacturer) module, industrial readers, and handheld readers. Fixed readers are divided into the base station reader and mobile reader. Fixed Readers are commonly used in the field of traffic control, OEM are used in fingerprint identification more, industrial readers are mainly used in the mining industry, animal husbandry, automated production and other industrial fields. Handheld readers are commonly used in the inspection and other non-fixed place. While containers are transported at the crossing of the inspection and, depot and terminal apron in the port, the label information may be updated at any time. Therefore fixed readers and handheld readers should be chosen. Base station readers of fixed readers are used to read the information of trailers and containers in the check bridge crossing, mobile readers are used to read the container numbers for quay cranes and yard cranes. Handheld terminals readers are used to read container numbers for the port mobile workers. Nowadays, RS-232/RS-485 communication interface of ISO/IEC18000 standard has been widely applied in the field of container transportation, so the reader designed in this article also select this interface.

Antenna Choice

The key considerations of tag antenna selection are the size of the antenna, the gain and the polarization. The gain of the antenna means the power density ratio of the actual antenna radiating element of the ideal signal generated at the same point in space premises under conditions equal to the input power, which quantitatively describes the extent of the centralized antenna radiation input power. The higher the gain, the smaller the transmit power required to achieve the same radiation effect. The polarization of antenna refers the direction of the electric field changes cause the change of the antenna identification direction while the antenna radiates electromagnetic waves to the space. Several commonly used tag antennas are shown in Table 3.

Table 3. Several Label Antenna

Antenna	Mode	Gain	Dimensions (wavelength units)	Impedance/Ω
Dipole	Omnidirectional	0～2dBi	0.5	50～80
Equivalent oscillator	Omnidirectional		0.5×0.05	100～300
Printed dipole	Directional	6dBi	0.5×0.5×0.1	50～100
Micro-strip antenna	Directional		0.5×0.5	30～100
Logarithmic spiral	Directional		0.5×0.25	50～100

From the Table3, Omni directional antenna has a peak gain of 0 to 2dBi.The directional antenna gain can be achieved 6dBi and energy consumes less contrast than the Omni directional antenna. At the same time, the polarization of the antenna includes linear polarization and circular polarization. The first three categories in the table are linearly polarized and last two categories are circularly polarized. Since the directionality of circularly polarized antenna is better than the linearly polarized antenna, so considering the gain recognition and directional perspective, the identification tag antenna should choose the micro-strip antenna or the logarithmic spiral antenna. Since the direction of tag antenna is not controllable, the tag antenna may also choose circularly polarized. Considering the panel antenna is applied more widely than Yogi-antenna and array antenna, so the reader antenna also chooses panel antenna.

Frequency Selection

A. Activation Frequency

Since the power supply capacity of tag battery is limited in active electronic tags, in order to reduce power consumption and extend its life. In addition to using the low-power components in the tag circuit, the most important thing is to rely on the wake-up mechanism. Active electronic tags are usually in the sleep state most of the time and transmit or receive data after they are woken-up. When data exchange is finished, the tags get back to the sleep state until the label be awakened once again. The reader antennas play the role of waking-up, only need to wake up the active tags around it. Because of the transmission characteristics of the electromagnetic wave, the energy attenuation degree of low frequency is lower than that of the high-frequency in the same distance and the low frequency has the character of the low absorption and strong penetrability for the metal material and liquid substance, so that the power of the tags is reduced to a lower clock frequency, which determines the low frequency has an absolute advantage to activate the tags. Low frequency 125 KHz and 134 KHz frequency do not conflict with other bands within the range of low frequency 135 KHz, not reserve for industrial, scientific and medical (ISM) frequency range; they have been used heavily by other radio institutions. The technology in this band is more mature and open. The transmitting power of 125 KHz desired is smaller compared with 134 KHz, therefore choose 125 KHz as the activation frequency of the active tag.

Table 4. Manufactures Conditions of Different Frequency Tag

Manufacture	Country	Label Type	Frequency	Applications
SAVI	USA	Active	433M, 2.4G	The United States national defense, logistics, port
INTERMEC	USA	Passive	433M, 900M	Wal-Mart, logistics
ALLEN	USA	Active / passive	433M, 900M	Geely, logistics
AMECH	USA	passive	900M, 2.4G, 5.8G	Highway
TI	Japan	Active / passive	2.4G, 5.8G	Toyota, highway
Toyota	Japan	Active / passive	5.8G	Highway
Combite	Sweden	Active / passive	5.8G	Highway
Tag Maser	Sweden	Active / passive	2.5G	Logistics, port, warehousing
Science and Technology Shanghai Rui Ting	China	Active	900M, 2.4G	Logistics, port, warehousing
Sichuan new source of modern	China	passive	5.8G	Traffic

B. Operating Frequency

Normal operating frequency is the basic technical parameters for radio frequency identification system. While choosing the frequency, national and regional standards should be considered besides considering the characteristics. Different countries and regions in the world may develop different standards, the EPC Global, the Ubiquitous ID Center (UID) and ISO / IEC three major systems are developed mainly in Europe and the United States, Japan, the International Organization for Standardization, respectively. Japanese electronic tags use 13.56MHz and 2.45GHz, Europe and the United States EPC standards adopt the UHF902 to 928MHz band. The information digits of Japanese electronic tags are 128 bits, but the information digits of EPC standard are 96 bits. In the application aspects, the UHF band in North America has been developed well, the active 2.45 GHz system is applied in Europe more and 5.8GHz system is applied maturely in Japan and Europe. Table 4 lists the frequency of major manufacturers of the world's different label production status.

From the Table 4, we can see the application of RFID frequency band is used in different countries in the field of traffic, port logistics, highway, and so on. Microwave segment has been accounted for a certain proportion and the majority usage is the application of 2.4GHz. With the increase of the international container trade year by year, the 2.4GHz band will be showing a broader prospect for the development and use of the container tag.

Because of the large container throughput in port, plentiful inbound and outbound trailer vehicles, fast moving and more high-altitude operations of loading and unloading machinery, so the label frequency of the RFID system should select those focused on large amounts of data processing and high transmission speed, long distance communication as reference. The frequency UHF 433MHz has been allocated to the amateur radio operators in the world, already belong to dedicated channel in China, the frequency UHF 860 to 960MHz are also belong to fixed and mobile services category in China, so the application of the above two types of band RFID can easily cause interference with other ISM equipment, the container tag frequency should focus on more than 2.4GHz band as the analysis object. In this paper, 2.4GHz operating frequency of electronic tags developed by Shanghai Port in Shanghai, China - United States, Savannah ship routes are referenced and take into account the characteristics that the electromagnetic energy transmission degree of attenuation is inversely proportional to the wavelength. They accord the international standards and select 2.45GHz frequency as the normal operating frequency of the electronic tag in container finally.

Communication Mode Selection

As mentioned above, there are inductor and the electromagnetic two coupled way, according to the working principle of radio frequency identification system. While in the inductive coupling, there are the tight coupling system and remote coupling system in the practical application, both of them use the contact space of radiate radio frequency channel to work between the reactive near-field inductive coupling of the RF tag and reader antenna (closed magnetic circuit). The operating frequency of the former is generally limited to 30MHz, the typical operating range is 0 to 1 centimeter; the typical operating frequency of the latter is 13.56MHz, working distance is less than 1 meter. In the electromagnetic coupling mode, the typical application is used in the long-range systems. It uses the non-contact space information to spread RF channel to work, which is formed by radio-frequency tag and the reader antenna radiation far-field region between the electromagnetic coupling. Their typical operating frequencies are 915MHz, 2.45GHz and 5.8GHz and their range can be up to 10 meters and beyond. So using electromagnetic coupling and long-range system can resolve the container data better, including the transmission and acquisition of information at a higher frequency, and also in accord with the characteristics of container port operations in the distance.

Work Mode

In full-duplex and half-duplex system, the RF tags response and send signals only after the reader issued the electromagnetic field or electromagnetic wave and the signal tags received is weak relatively the signal emitted by the reader. So the load reflection modulation techniques must be used to load the tag data to reflection echo, commonly used in the passive tag system. In contrast, the timing system application uses radar mode of operation. According to Maxwell's electromagnetic field theory, the variation of the electric field will generate the magnetic field around it, and the changing magnetic field will also produce the electric field. In the timing system the electromagnetic fields generated by the reader radiation are in the periodic variation, the tags will identify the electromagnetic fields and use them to transmit data from tags to the reader. By contrast, we can see the timing system need not process the data modulation so can save a lot of energy. The large amounts of data need to be processed in the container transportation, low power factor should be taken full account and therefore, the timing mode may be the ideal work mode of container tags.

Modulation Pattern Selection

The modulation pattern is the most important aspect of the RFID system communication and almost affects the performance of the whole system.

Generally, the reader of the port can read 20 tags per second, which can meet the practical requirements in the application process. If each tag contains about the amount of 1kb data, then the total read data rate should be 20kb/s and belong to the category of low-rate communication. So there is no need to use high modulation efficiency way can meet the practical requirements in the container port. Considering the emission power limitation of 2.45GHz band, direct sequence spread spectrum pattern is selected as the best choice which can also meet the low-power design requirements.

Integrity and Security Protection of Data Transmission

In order to guarantee container data can be read, transmitted and received correctly at the multi-tab environment in port, avoid the occurrence of missing read, misread, out-of-order and data collision, effective error control and anti-collision function should be taken in the process of data transmission. Currently, the technology has been used better to solute these problems, such as microwave communication error detection, Cyclic Redundancy Check, binary search algorithm and so on. Container transport through many links, expensive goods in containers have the possibility of being illegally stolen, tampered and attacked. Therefore actions should be taken to the tag information, such as encryption and authentication, only the personnel granted permission have the right to obtain information of the goods.

Equipment Installation

A. Electronic Tags Installation

Installations must not affect the line-of-sight regional of the driver, the trailer electronic tag should be installed at the bottom of the rear-view mirror on the inside of the front windshield of the driving indoor. The environment and the reliability of acquiring the RFID tag data by various equipment operations of the Container Terminal operations must be taken into account, such as the crossing, quay crane, RTG, fork lift truck and so on. Tags cannot be crushed by other containers or slings for lifting loads, so electronic tags in containers should be installed in the middle of the container door lintel.

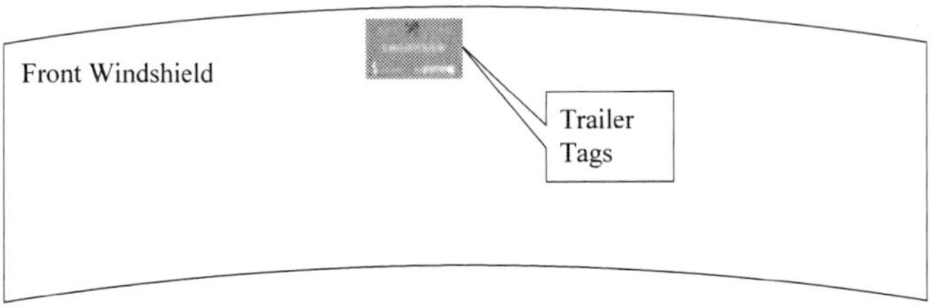

Figure 6. Install Position of Trailer Tags.

B. RFID Equipment Installation

1) Crossing Equipment Installation

In order to ensure the information about the trailers and single or dual-containers with tags be collected at the same time when passing the crossing channel, and prevent the interference between adjacent channels, two top-loading antennas are designed to set in front and behind, and the distance interval of two antennas is 14 meters or so. While the trailers stop above the weighbridge, the distance from the antennas to the front and rear door horizontal direction are just about 1 meter respectively. This ensures that the antennas can receive the information of trailers tags and container label. The reader is mounted in the middle position of the isolated island in the weighbridge. The installed location of crossing is shown in Figure 7.

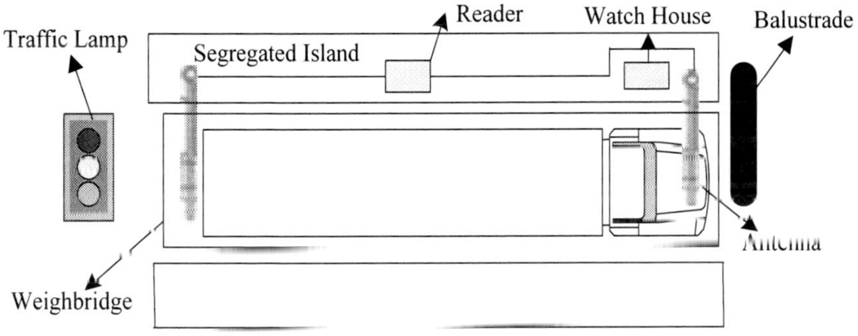

Figure 7. Installed Location at the Crossing.

2) Installation of Equipment on the Quay Cranes and Gantry Cranes

While the quay cranes and gantry cranes operate, spreaders of cranes grab the top surface of the container in perpendicular direction. Spreaders can take flexible adjusted according to the container size (20 feet or 40 feet).So, the reader antenna is installed in the bottom of the both sides of the spreader frame in the downward direction. Taking into account the container door may toward front or behind, two antennas are designed to set in the spreader to ensure that the container label information can be collected in the program. The reader is installed in the middle of the spreader, so can take power easily from the middle of the spreader power junction box. Installation on the quay cranes and gantry cranes is shown on Figure 8.

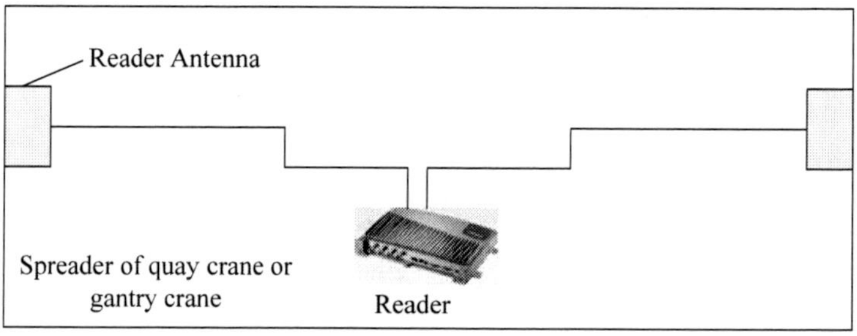

Figure 8. Installed Location on Crane.

3) Installation on Front Lifting

Due to the space between the beam across the bottom surface of the front lifting spreader and the container top surface is only 1.5 centimeters, the reader antenna can not be installed on the spreader bottom, which can only be installed in the side of both ends of the spreader beam. So two antennas are set at both ends of the spreader to ensure that the tags on the container can be collected, the reader is placed in the intermediate position of spreader to facilitate the power supply. Installation on the front lifting is shown in Figure 9.

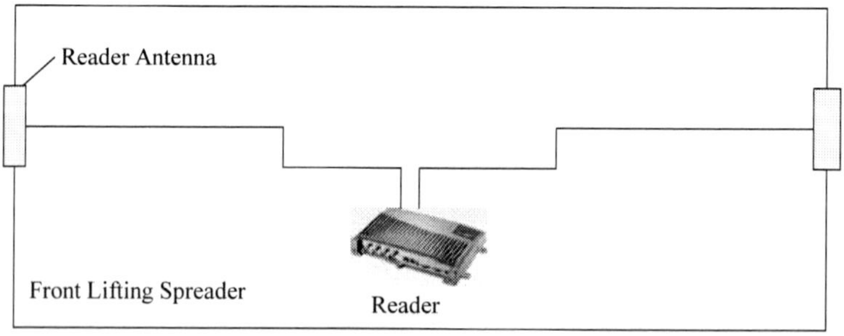

Figure 9. Install Location on Front Lifting.

4) Installation on Stacker

While the stackers operate, empty containers are lifted by the bayonet lock of the two arms cached on the two holes in the side of the container top, so the antenna is mounted on two grasping arms on the container top and slightly downward sloping and dual antennas are set at the same. Taking into account the vibration and shock of the stacker while operating, the reader is installed in the horizontal beam of the two grasping arms.

Equipment installation scheme considers the characteristics of the machine operations and the convenience and RFID technology. The program is only for the general case and if encounter some special circumstances, special treatments should be made and handled.

Electronic Tags System Construction, Application and Evaluation in Container Port Based on RFID Technology

Container Port Electronic Tag System Construction

Structure an intelligent information platform and reach digitized in container port are the development requirements of the advanced productive forces in today's world shipping. They are also the important symbol of the modernization process in the container port and the important criterion to measure the management services level in the container terminal. Therefore, promoting the digital construction of the container port has the realistic and far-reaching impact to enhance its ability to participate in international competition.

A. Overall System Structure

Aiming at the container shipping information processes and the container terminal operating process, the port information platform is established based on RFID technology. It not only optimize the organization of port operations, improve operational efficiency, integrate the port EDI information at the same time, but also provide real-time information inquiry service in the container logistics aspects. It laid the foundation for the integrated management in the realization of the international container real-time online monitoring of the whole transport process.

According to the flow of information process, container port information platform based on RFID technology can be divided into four layers, they are physical layer, communication network layer, data transmission layer, application layer, as is shown in Figure 10.

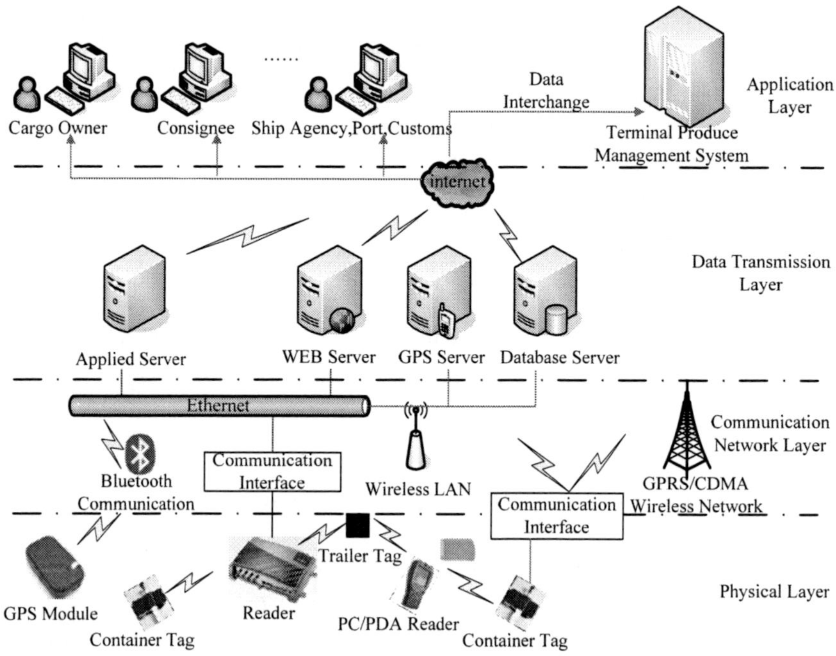

Figure 10. Overall Structure of the Container Port Information Platform Based on RFID.

(1) Electronic Tags Automatic Identification System

The automatic identification system of electronic tags belongs to the physical layer and mainly completes the real-time collection and automatic identification of the vehicle information and container information. It guarantees the data information about vehicles and containers can be captured and recognized automatically, quickly and reliably. Identify ways use non-contact technology which can identify and read and write the moving and stationary tags of trailers and containers. The identification system includes the vehicles electronic tags, container tags, fixed readers, handheld readers, GPS module and so on. Using fixed reader and electronic tags can check the label information collection of bridge, yard, and quay apron, etc. Using a handheld reader and electronic tags can be applied to the label information which needs to change the manual entry in the non-fixed location. Such as vehicle violation information input, yard temperature control box temperature entry, and terminal apron tally entry. The GPS module is used to track the positioning trailers and the geographical coordinates of the container.

(2) Hybrid Network Data Transmission System

The wired network, wireless communication network and the LAN in port can be integrated in the communications network layer which can provide flexible network access. Data can be analyzed, stored by server in transport layer so real-time data of electronic tags can be uploaded and processed to ensure that data transfer of the containers has not blind area of the port and other transport arenas.

(3) Container Port EDI and Business Inquiry System Based on RFID Technology

EDI "one-stop" services and terminal business inquiry system needs to be opened in order to meet the requirements of Customs and official inspection regulatory for containers and transport vehicles, dynamic understanding the containers and transport vehicles for the ship, cargo and the port side so the container transport participants can understand the current situation at the first time.

B. Modular Design and Function of the System

(1) System Modular Design

According to the work flow of container logistics and operating characteristics in port, the basic functions of the electronic tag system can be determined in container port based on RFID technology. The electronic tag system can both solve container port node bottlenecks in the production process but also meet the need for customers to keep abreast of the information about the goods and further improve the functions of the real-time monitoring of the import and export goods from the customs and other official inspection departments. The system module is shown in Figure 11.

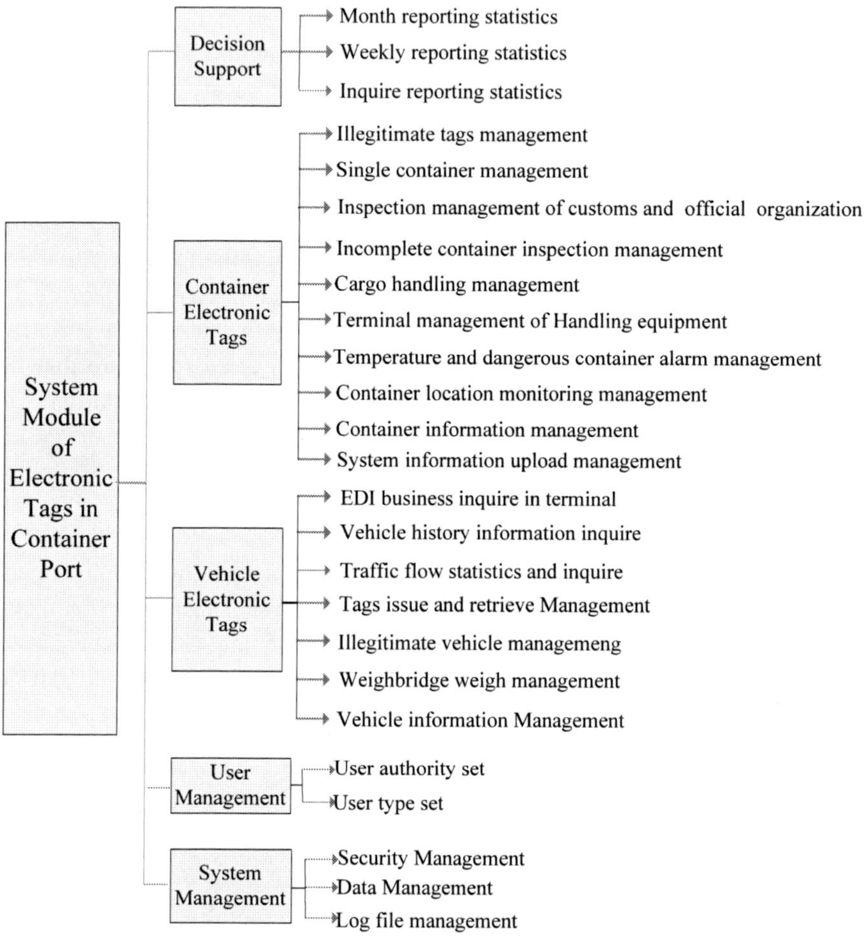

Figure11. Electronic Tag System Module of Container Port Based on RFID.

(2) System Functions of Electronic Tags

Directed by the concept of modern logistics management, using RFID-techniques, combined with GPS, smart terminal, EDI and other digital and information technology, transportation, management, loading and unloading activities are monitored in the port logistics process. The information collection and automatic identification can be realized at entry and exit crossing, the yard, vessel handling and other aspects in container port. EDI and business query system were established in container ports based on RFID technology which can trace the full information of container logistics process, grasp the full cargo information of container logistics in real time, realize the intelligent management of container ports and visualization during transportation. In order to improve port operations efficiency and reduce management costs. The operation of the system can strengthen the management of the containers and trailer vehicles in container port, and will further strengthen the supervision of the customs, and improve the efficiency of the Customs Department. Therefore the electronic improve the overall modern management level and service level of container port logistics, reduce logistics costs. The system can achieve the following functions in container port crossing, yard, quay cranes and customs inspection.

1) Loading Site

After the shippers complete binning and close the door, they can use a handheld reader to scan the electronic tag, and then the electronic tags in containers are bound with the container. After binding, containerized cargo information can be inputted manually in the control software. The control software interaction via Bluetooth and GPS modules, get the current GPS coordinates and analyzed by the GPS server, containerized cargo information and GPS place names are sent to the reader by the control software. After writing data to the electronic tag of the containers, the electronic tags are installed in the container door header. While for the trailer tags, vehicle information can be inputted into tags by port authorities when trailers are transected pass in and out registration, tags are attached to the front windshield of the vehicles and cannot be privately removed. Finally containers are transported to the loading pre-entry point of the port by trailers.

2) Port Crossing

The information about trailers and containers are inputted to the container port electronic tag system at loading pre-entry point, then are sent to the port production management system via EDI center. When the trailers enter the load crossing on port, the electronic tags of trailers and containers go through the coverage area of the crossing fixed reader, the readers receive signals from electronic tags, read the information in the electronic tags and upload to the application server in real time. Control software of the application server automatically compares and matches electronic tag information, and determines whether the tags are legitimate. If the verification is correct, the system will automatically print driving directions, and then the trailers drive to yard load or unload containers. If the information is incorrect or container is illegally opened, the system gives an alarm, and the container is inspected by the customs.

3) Customs Inspection

When the customs need to inspect the containers, the electronic tags are corresponding authorized and reopened by the reader and running the control software through PDA, at the same time the data are transferred to the background management system. Customs open the electronic tags legitimately, after inspection the reader head completes the corresponding authorization sealed to the electronic tag, again transfer relevant data to the background database, and finally hung up the electronic tag.

4) Loading Yard

When handling the machines of the yard (such as gantry cranes, stackers, forklifts, and so on) carry out the loading and unloading operations, the readers installed in the machinery and equipment at terminal automatically read the information on container tags, if be verified correct, the driver can implement the lifting operation, otherwise the system alarm.

5) Loading Quay Crane

When the vessels load containers on board, the reader installed on the quay crane can automatically read the container tag information, if be verified correct, and then shipped, otherwise the system alarm.

6) Unloading Quay Crane

When the vessels unload containers, the readers installed on the quay crane automatically read the container label information, if being verified correct, then the trailers will unload containers and transported to the yard storage. If the information is incorrect or the container is illegal, system alarm open, the inspection will be executed by the destination port customs.

7) Discharge Yard

When the discharge yard machinery (such as gantry crane, stacker machine, forklift, and so on) execute loading and unloading operations, the reader installed in the terminal of the equipment automatically read the container tags information, if be verified correct, and then the driver implements the lifting operation, otherwise the system alarm.

8) Discharge Crossing

While the trailer load container out the yard, the electronic tags of vehicle and container go through the coverage area of the reader, the information in the electronic tags are automatically read, if be verified correct, and then the trailers go to the unpacking site. Otherwise, the system alarm.

9) Unpacking Site

After the consignees confirm the status of tags, PDA / PC authorized to remove the electronic tags, tag data are empty and tags are retrieved.

(3) Functional Design of Each Module

1) System Management

Real-time tracking and recording system access log file is the basic function of the system, so administrators can quickly record and predict potential system invasion, prevent malicious distort and hacker attacks, ensure the normal operation of the system. The data collected can be carried out effectively storing, analysis and processing by the system, so to guarantee the independence, reliability, security and integrity of the system data, reduce the data redundancy, improve the data sharing extent and the efficiency of data management. If the tags are illegally destructive, the system can send alarm signal automatically. The functional design of the system management module is shown in Table 5.

Table 5. Function Design of System Management Module

Management contents	Realized Functions
Log Management	Real-time tracing, recording system access log file, prevent malicious distort and hacker attacks
Data Management	Effective storage, analysis and processing of the data
Security Management	Provide the automatically alert service for containers

2) User Management

The system can set a different logon password for different types of users, assign appropriate permissions for users so that each user entering platform faces different user interface, users can only use and manage the information within their duty range.

Ports are the important logistics nodes in the container transportation. The trailer vehicles should accomplish loading and unloading operations in port and containers are transferred in port too, so the port should have the overall management of the container information. But the operations of customs and other inspection belong to the functions of the official inspection management. Cooperation and collaboration are just playing in the identification process in container ports and the customs and inspection experience job manage permissions have not included the process of using the system. Before the Cargo and the freight forwarders organize the trailers entering the port, the information about trailers and containers need to input by customer's terminal or port pre-entry point, so the cargo owner and the freight forwarders should have the permissions of trailers information management and container information management besides the public information search platform. For shipping companies and shipping agency, they are the main actors of international container multimodal transport and their most important duties are carrier at sea. They do not have a close relationship with the transport trailers and goods inside the container, so only open "EDI and terminal business query" permissions can meet the requirements. The user management module functional design is shown in Table 6.

Table 6. Function Design of User Management Module

User Type	Permissions
Port	System management, vehicle tags management, Container tags management (not including customs, inspection experience operations management) and decision support
Customs, Official inspection	EDI and terminal business inquiries, customs, official inspection experience in operations management
Cargo, freight forwarder	EDI and terminal business queries, trailer and other vehicle information management, container management
Shipping companies, shipping agency	EDI and terminal business inquiries

3) Trailers Electronic Tags

While entering into the port, the electronic tags of trailer storage the job number, license, plate number, vehicle type, company and other basic information.

The intelligent weighing can be reached by RFID technology at the bridge crossing in the terminal.

The yard administrator can handle illegal vehicles on-site operation of punishment using a handheld terminal reader and punishment information can be inputted to trailer tag and automatically upload to EDI center and terminal business systems. These realize the management of port punishment in one place.

Historical operating records of the trailers and the traffic flow in crossing can be queried by the system operator, and the case of traffic jams can be eliminated by dynamic traffic induction in real-time. Trailers tag function module design is shown in Table 7.

4) Electronic Tags on Containers

The electronic tags of containers can store the name of the carrying vessel, voyage number, the port of loading, transshipment port, the port of discharge, container number, container type, bill of lading number, seal number, container state, get in and out signs, domestic and foreign trade signs and other information. When the handling the equipment to install the electronic tag reader, the driver can quickly find the container and identify it according to the radio frequency signal. The containers in the yard can be displayed in real time via GPS satellite navigation exact location and the position monitoring allows yard graphical visualization and determine the stacking position is correct or not, the mechanical drivers in the yard can pick up and select a container placed position by touching the terminal, reducing the status of the yard machinery temporary turnover container still need to be corrected. Tags contains temperature and humidity alarm, it is connected with the sensor inside the container, can automatically monitor and alarm to the temperature control containers and dangerous goods containers in real-time. The over-landed, short landed and damaged containers can be record through the real-time terminal reader entry by the terminal tally clerk on board, so the process of paper-based documents can be reduced. Customs and other official institutions can open and inspect container with a handheld terminal, and input the switch containers records to the tag and implement effective supervision to the container cargo. When the container tags are scanned by the crossing and go into the port, the container information can be automatically uploaded to the EDI center and terminal business inquiry system, so customers can understand the state of containers at any time. Function module of the container tag design is shown in Table 8.

Table 7. Function Design of Trailer Tag Module

Vehicle tags manage content	Function
Vehicle information management	Record information of trailer operation number, license plate number, and owned motorcade, etc.
Weighbridge management	Automatic weighing trailers and containers weight
Illegally vehicle management	Suspend illegitimate trailers, including suspension start and end time, the cause of the suspension
Traffic flow count and query	Count the number of empty and heavy vehicles in the yard, query the number of empty and load vehicles at crossing per unit interval
Vehicle history information inquiry	Query the history of trailers get in and out of the port and the number of illegal operations
Tags issuance and recycling management	Registration and cancellation of the tags
EDI center and Terminal Business Inquiry System upload information management	Irregularities vehicle information is uploaded to the EDI center and terminal business query system, if there is a suspension, it is disabled everywhere, so realize the unified management in all ports

Table 8. Function Design of Container Tag Module

Container tag management content	Function
Container information management	Input the name of the vessel, voyage number, container number, bill of lading number, loading port, destination port, container type and other information
Container position monitoring and management	Display the exact location of the yard container in real time using GPS satellite navigation
Temperature control and dangerous container's real-time alarm management	Monitoring and management can be made to the temperature control containers and dangerous goods containers by temperature sensor on the tag, if the temperature inside the container is abnormal, it can automatic alarm
Wharf handling equipment terminal management	When the container tags are read by crossing reader, container number information will be automatically displayed on the loading and unloading equipment terminal, when the container transfer operator can touch to confirm, which can reduce the operational link of walkie-talkie
Tally management	The wharf apron tally clerk can implement cargo handling operations by the PDA terminal, reducing the links of walkie-talkie and paper records, quay crane drivers can operate without the tally clerk job
Inspect damage management	The container inspector at Crossing can implement inspection and records to the damaged container by the PDA terminal, without paper records
Customs and official inspection operations management	Labels can automatically record the time, place and number of opening the door authorized Customs and inspector staff to unlock and inspect the labels by PDA terminal
Single-container information inquiry	The name of the vessel, voyage number, container number, bill of lading number, container type, destination port can be queried
Tags exception management	Update operations management can be implemented on the label in damaged case
EDI center and terminal business inquiry system upload information management	The name of the vessel, voyage number, container number, bill of lading number, destination port, container type, state for transport, Customs, docks release and other information can be queried.

5) Decision Support

The terminal management department counts inbound and outbound trailers traffic and container volume under the electronic tag system, print the report form and analysis the development of the port, which provides the reference and guide for the follow-up development.

C. Data Transmission and Exchange Process in System

The electronic label of the container port management system based on RFID technology includes the electronic tag identification system, container electronic tag identification system and the operate terminal system.

Supported by the station database management system, take a wireless data transmission system as the carrier, accurate and real-time acquisition and transmission are implemented to the inbound transport vehicles and contains information.

The history and current state of transport vehicles and container can be queried, then get in and out of the crossing operating record can be counted, yard container storage location and the current storage state can be recorded and queried, so to achieve standardized management of transport vehicles, automation management in crossing the yard, documents paperless operation and transparent management of container supply chain information in real-time.

The transmission and exchange process of the system data is shown in Figure12.

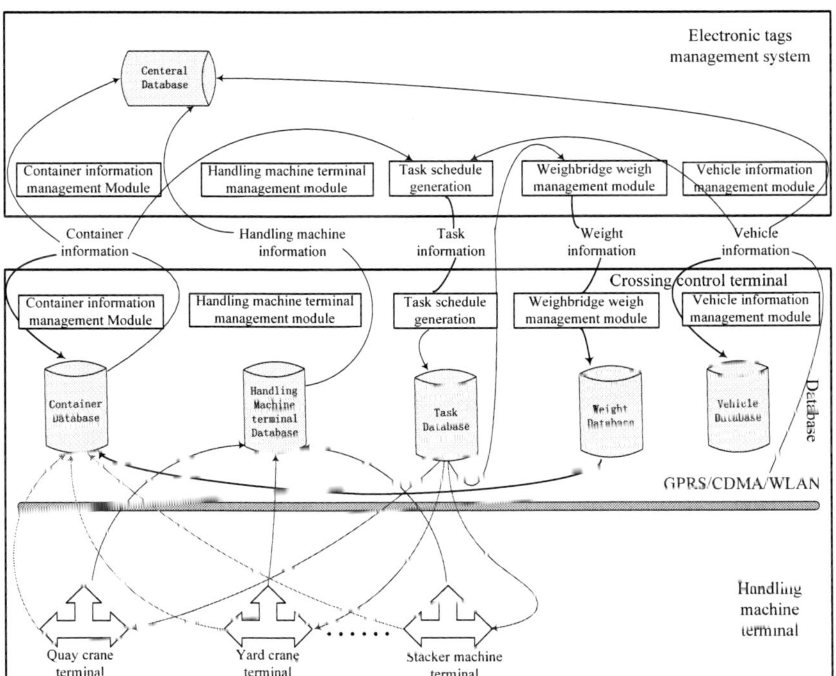

Figure 12. RFID data flow between database management systems and data exchange.

The specific process is as follows:

Before the trailers enter the port, the trailers and income/ pickup container information are inputted respectively to the electronic tags of vehicles and containers by the freight forwarding or container entry point through the container ports electronic tag system. Vehicles and the label information of containers are used together to generate the inward port task scheduler. After the trailers have access to a terminal crossing, the task information about entering the port is read and stored in the task database by the crossing control terminal. At the same time, the container information and the trailer vehicle information are stored in the container database and the vehicle database respectively.

The task database directs the weighbridge to weigh, and sends the weight information to the container database by the weight database transmission, automatically matches and upload the actual weight.

After the handling equipment terminal displays the task instructions, the container is picked up, the information of containers and equipments are automatically uploaded to the container database and the device information database and ultimately stored in the terminal central database. While the trailers depart through the crossing and be scanned, vehicle database automatically uploads trailer information to the terminal central database.

Application of the Electronic Tag System in Container Port

A. Application between Ports

RFID tags combined with existing EDI information, which realizes the information collection and automatic identification of container transport between domestic and foreign ports and provides a convenient service for shippers. Freight forwarding and shipping agency meets the requirements of customs supervision. The specific application process is as follows.

After the shippers and freight forwarding complete loading and close the container door, the tag is authorized and hanged. The information of Entry Vessel, Voyage, container number and other are inputted in the tag. The trailer transport the container to the pre-entry site of the port, the information about vehicles and container are inputted and uploaded by EDI to the terminal vehicles and container label system. Then the trailers get into the port and begin loading. The reader devices installed on pier crossing automatically read the vehicles and container labels, if confirmed correct, trailers go into the port. In the process of container entered into the yard and shipping, the readers installed in the device under spreader read container number information, containers are loaded on board and departure. The container label information is sent to ship agency at the charging port via EDI, and then sent to ship agency in discharging port by ship agency of the loading port. Finally, the shipping agency of discharge port sends information to the terminal container label system in discharging port. After the ships arrive the port of discharge pier, the containers are unloaded and stockpiled in rear yard under the correct reading of quay crane yard and yard equipment. The trailers enter the port and pick up the container after delivering container stuffing charge, march out by crossing identification. At last, the label is recovered, the consignee pick up the goods. The application process is shown in Figure 13.

B. Application on Port Internal

(1) Containers Enter Port Processes

Trailer drivers carry the container to the port pre-entry site, the containers and trailers information are typed in and formed the EDI form sent to the system of electronic tags at terminals. When the driver drives to the crossing, the container inspector test damage through handheld PDA terminal. The vehicles and container number are extracted, matched and upload by RFID equipment and the system assign a container yard location and send voice prompts. The drivers print driving directions according the voice prompts, go to the specified container field bit and wait for unloading containers. The drivers of yard machinery use the RFID reader to identify the container numbers according to the terminal instruction, lift up the containers to yard and confirm at the terminal. After the containers come into the yard, the storage yard instructor uses handheld PDA terminal to record and trace the situation of the temperature control and dangerous containers. The specific application is shown in Figure 14.

(2) Containers Leaving the Port Process

Container trailer drivers drive to the port pre-entry site holding the lading bill, the container and vehicle information are inputted and send to port electronic tag system in the form of EDI.

When the drivers drive to the crossing, the RFID device reads the vehicle label, the system automatically obtains the container number mentioned, according to the correspondence of the vehicles and containers and sends voice prompts.

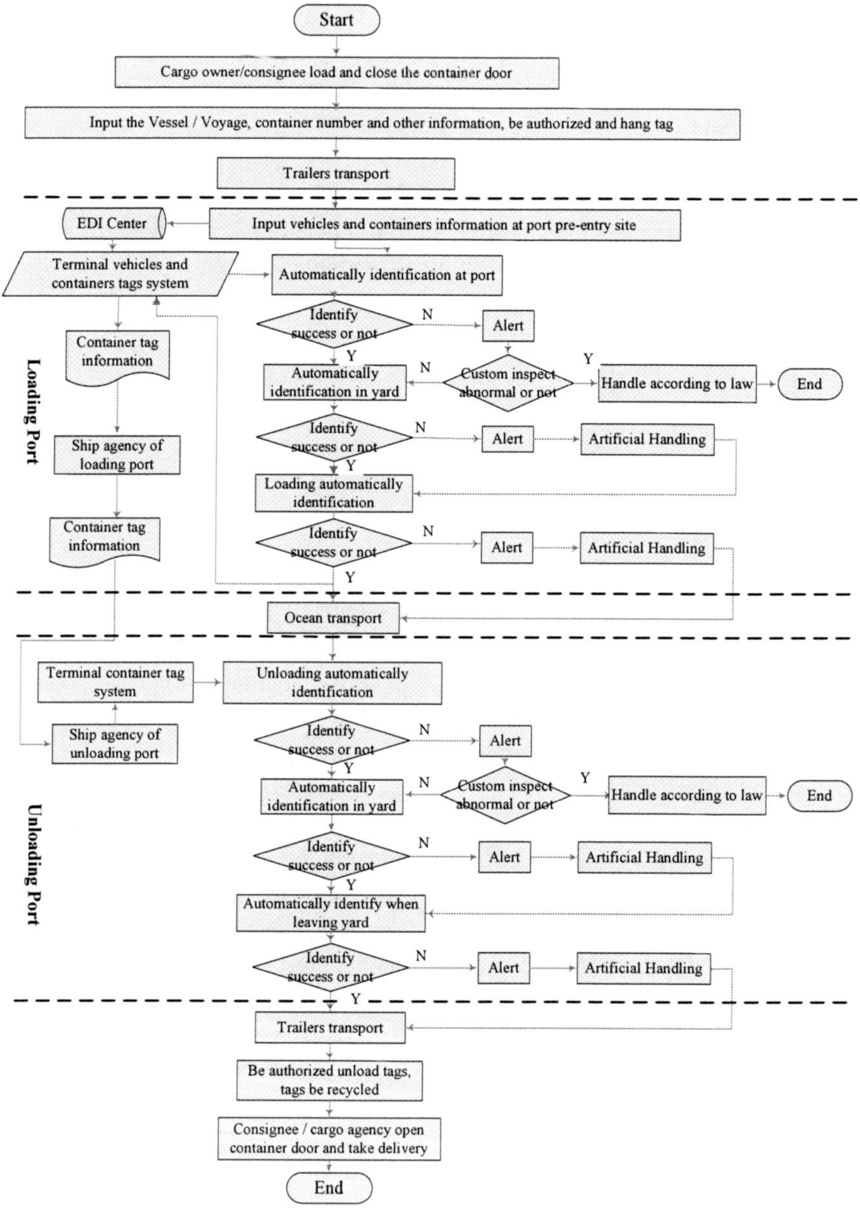

Figure 13. Application Process of Container Electronic Tag between Ports.

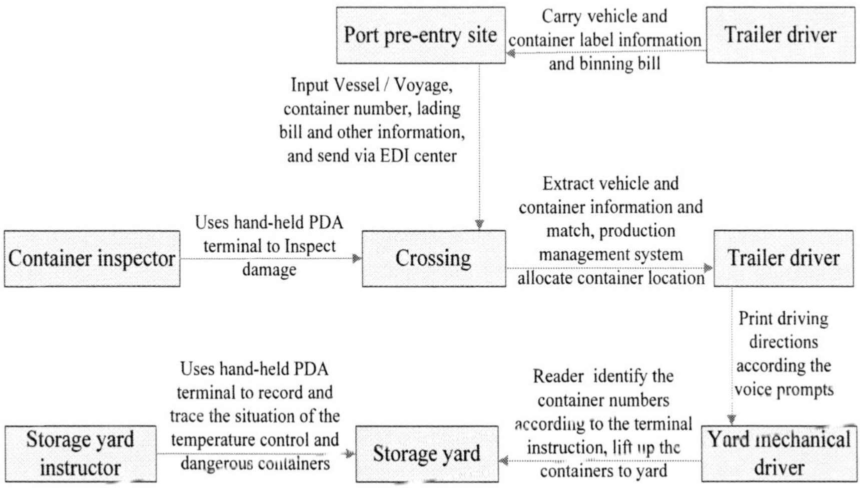

Figure 14. Process of Container Entering Port.

The drivers follow the voice prompts to print driving directions and go to the yard and pick up the containers. The yard machinery drivers use radio frequency to identify the container number, and then hang up containers to trailers. Container trailer drivers accept vehicle and container labels information collection at crossing, drive the containers leaving port area. The specific application is shown in Figure 15.

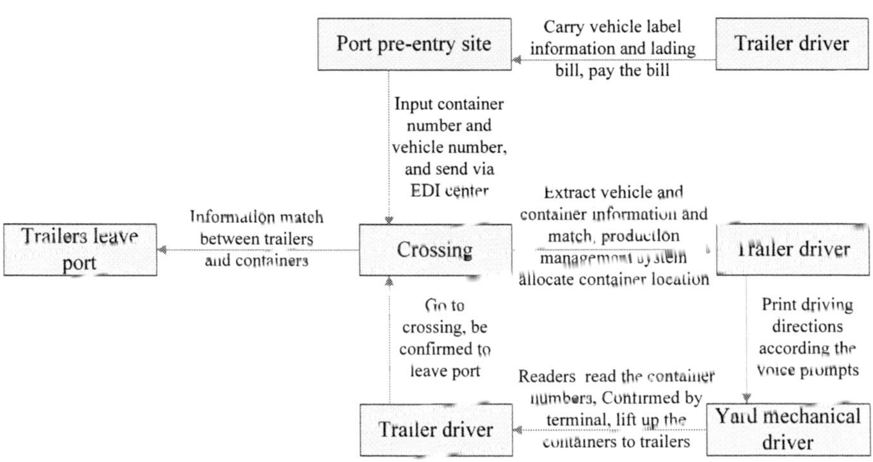

Figure 15. Process of Containers Leave Port.

(3) Export Shipment Process

After the reception center completes the export container permitted through, receives the cargo plan and manifest, notifies the yard to statistics, container volume, makes loading bay plan and the control center allocates tasks for the handling machinery. While loading on board, the yard mechanical drivers use the reader to read container number according to terminal commands, hang the containers to the trailers, trailers carry the containers to the quay crane for shipment.

The quay crane drivers begin loading for the ship with the terminal display of the vessel and shipment container number. The tally clerk holds PDA terminal to record over-landed, short landed, and the damaged container in real-time, as is shown in Figure 16.

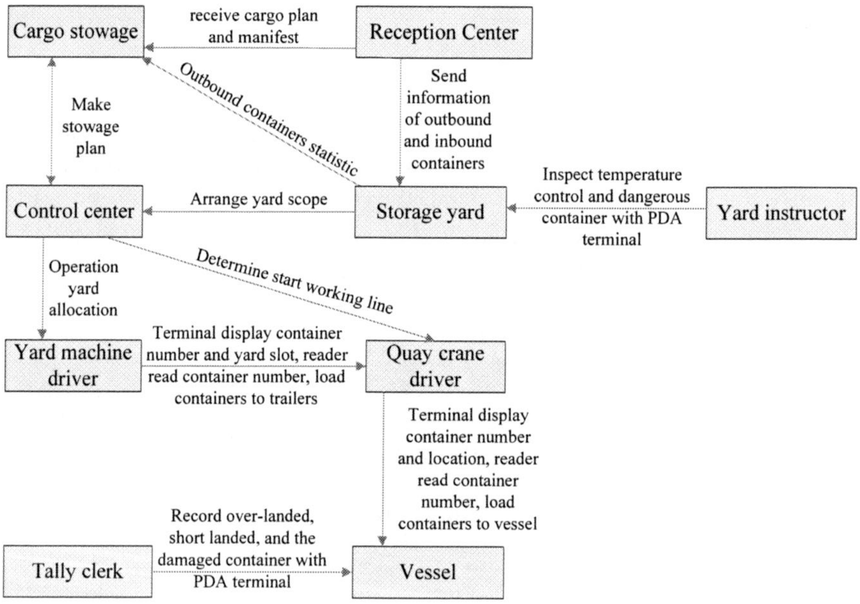

Figure 16. Loading Process of Container.

(4) Import Unloading Process

After the Reception Center receives bay plan and manifests, notifies the yard to arrange stowage venues and unloading sequence, the control center allocates tasks to the handling machinery. While unloading from the ship, the containers are also read by the reader of quay crane and yard machine, transported by the trailers in the yard in horizontal delivery. When the containers are in the unloading course and unloaded in the yard, the tally clerks and yard instructors hold PDA terminal to input the over landed, short landed, the damaging information and monitor the temperature, as is shown in Figure 17.

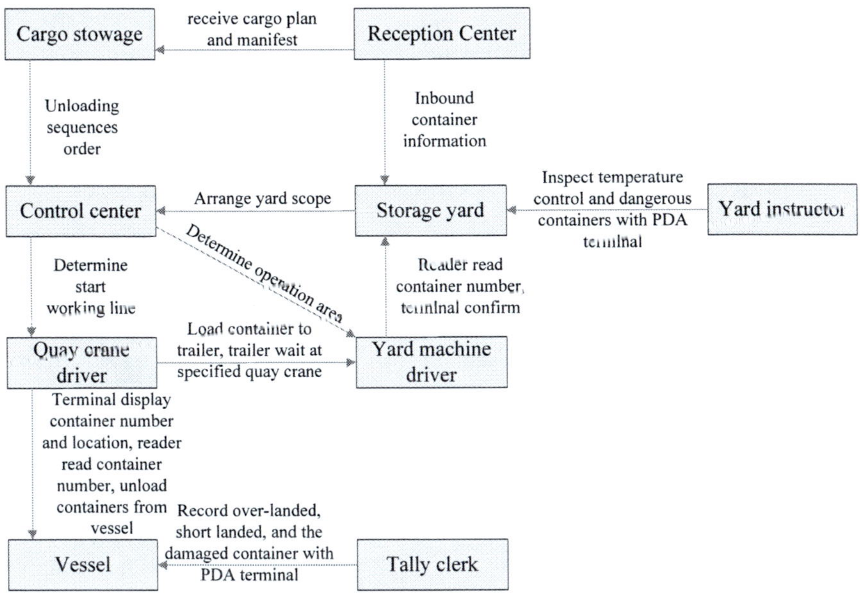

Figure 17. Unloading Process of Import Container.

Evaluation of RFID Electronic Label System Application in Container Transportation

Reduce Error Rate, Mitigate Artificial Labor Intensity

Compared with the application of traditional operating way in container ports, the electronic tag system can help the port to realize the management of trailer vehicles and containers at the same time. At the crossing, the two labels can carry out information collection and automatic identification to the trailers and the containers. It excludes the manual entry point, the recognition rate of more than 99% effectively suppresses the 35% error rate in hand container bill entry. In the storage yard, the reader installed under the spreader of gantry crane and other mechanical equipment can automatically identify the container number, so alleviate the burden of the driver using the naked eye to identify the job thereby eliminating the driver's fatigue. On the side of the vessels, the quay crane drivers manipulate the reader under spreader carrying out long-distance automatic identification of container number in the altitude of more than 50 meters, without the cooperation of the tally clerk. At the same time, the crossing container inspector, container yard instructors, and the vessel tally clerk all use the PDA terminal to carry out temporary maintenance on the entry and exit information of trailers and containers instead of using traditional paper records. The contrast of the situation of the system applications before and after is shown in Table 9.

Table 9. Application of System Comparison Before and After

Place	Before applying	After application
Crossing	Trailers and containers information with an error rate of 35% compared with manual entry and paper records, serious abuse of a trailer license number	Automatic identification and the PDA terminal maintenance trailers and containers, recognition rate can reach 99%, effectively combat trailers license abuse
Storage Yard	Mechanical drivers contact central control with a walkie-talkie, identify the container number visually, feel fatigue easily, instructor 24-hour patrol temperature control and dangerous containers record on paper.	Reader of the spreader automatically identifies the container number, do without walkie-talkie, the naked eye only needs to assist, when the temperature is abnormal, labels automatic alarm, instructor maintenance using PDA terminal
Vessel Side	Quay crane drivers need cooperation with the tally clerk by walkie-talkie, the tally clerk paper records over landed, short landed, damaged container information in the paper	Readers under the spreader carry on a long-range identifying container number at 50 meters high-altitude, quay crane drivers can operate all alone; the tally clerk maintenances over landed, short landed and damaged containers with PDA terminal

Strengthening Customs Supervision and Improving Transport Information Sharing and Transparency

Since the electronic tags of container have the function of automatically alarm if opened illegally, so the reader can determine whether the label is legitimate when reading container number at the crossing before exports or the unloading quay crane before the imports. If the containers have been illegally opened, the smuggling and dangerous goods were stored or the illegal border crossers were sneaked into the container in transit process, the labels can alarm promptly. The officers can quickly dispatch to handle, which overcomes the drawbacks of only after the containers go into the yard can be inspected in traditional mode and ensure the safety of port operations. The customs officers may use a handheld PDA to inspect, compared with the traditional mechanical inspection thereby time and effort are all saved. The tag information is uploaded to the EDI center and the port business inquiry system. So, the customers can keep track of the container transport information. In view of this point, the application of electronic tags in container transport can reduce the stolen goods, strengthen customs supervision, improve the customs inspection speed, and enhance container transport information sharing and transparency.

Improve Port Capacity

Container port electronic label system has significant benefits for speeding up trailers entry and exit the crossing, increasing port container throughput.

(1) Speed up Trailers Crossing through Speed

RFID tags scan time in about 5ms to 5s, which is much faster than recognition speed of other technologies, and the RFID reader can read multiple tags once, which greatly improve the efficiency of goods identification. The RFID technology is applied in the container port, reflecting its rapid identification advantage.

Take the Dalian Port container number identification based on the trailer identification of the RFID technology and optical technology, for example, to illustrate the application of the electronic tag system in container port can greatly facilitate the crossing ability. Survey and count the crossing territory of two empty and three heavy vehicles departing according to the Dalian Port certain period time (such as 1 week).The recognition rate of trailer number is above 99%, the average container number recognition rate is more than 95% and the mean transit time of empty vehicles going through the territory crossing within 30 seconds is more than 95%.The mean transit time of heavy vehicles crossing departures within 30 seconds is also more than 95%.The performance of intelligent crossing comparing with traditional crossing is shown in Table 10.

It can be seen that the trailers identification and container number identification using RFID technology make the container port crossing through capacity been greatly improved and the processing time is reduced to about half than before. The radio frequency identification rate of the RFID technology is higher 4% than the image recognition of the optical technology rate, and the identification time is short. Therefore, the container ports adopts the dual-label system of RFID technology will be even more enhance the crossing through capacity and speed up the trailers' passing speed at crossing.

Table 10. Performance Comparison between Intelligent and Traditional Crossing

Performance Comparison		Traditional Crossing	Intelligent Crossing
Process Performance at Normal Case	Inward heavy trailer pass crossing	90～180 s	60 s
	Inward empty trailer pass crossing	90～300 s	30 s
	Outward heavy trailer pass crossing	60 s	30 s
	Outward empty trailer pass crossing	30 s	15 s
Processing Speed	Trailers identification	Artificial input	<0.5 s
	Container identification	Artificial input	<5 s

(2) Improve the Throughput of the Port

Suppose each container can speed up the processing speed for half an hour at land links, then there are over 1.14 million TEU additional throughput increase per year in Chinese port. If the RFID tag system in container port is implemented, then the actual time of every container in the land will be saved far more than half an hour. Assuming that each international container from the export terminal flow to an import terminal need eight weeks, use RFID technology in part to transform container transportation management, it will greatly improve the container port throughput, but also bring more and greater development potential and development of container transportation.

Label Cost

Using RFID technology to transform the information construction in container ports will indeed bring the improvement and the cost of the container shipping industry on a certain extent but the improvement of the efficiency in container logistics can completely make up the raising cost caused by the introduction of new technology. The transformation of technology will not be a threat to the parties in the industry but will attract all parties to have plunged in and use the RFID technology. In recent years, with the rapid growth of port container throughput, the need of the electronic tags for trailers and containers will also show an increasing trend. However, from the point view of nowadays situation, the cost of the electronic tags from dozens of U.S. dollars in the past drops to several U.S. dollars, and the cost has continued have the trend of decline. The proportion of several dollars investment for each container is smaller relative to construction cost (a 20-foot container cost 2 000 $, the 40-foot container cost 3 000 $). From the long-term development angle, it can be accepted in the container transport industry as a whole point of view.

RFID Application Analyses in Container Port

Nowadays, the RFID technology has many applications in container ports both at home and abroad, the following is the concrete analysis.

(1) Los Angeles Port Introduce RFID Technology for Management

The Southern California terminal of Los Angeles Port introduces the innovative RFID technology, collect the inbound and outbound container information automatically in real-time and improve the speed and accuracy of containers handling at the terminal. All containers get in and out of the terminal are transported by the trailers and the system can immediately identify containers affixed with RFID electronic tags on trailers and confirm its position. Through logistics information based on RFID technology, the port cargo pass rate has been greatly improved in Los Angeles Port.

(2) Shanghai Port Container RFID System

In 2005, container tag industrial experiment of Chinese domestic Shanghai Port to Yantai Port opened the Chinese container tag application precedent. Container tag used is active and using the unique activation frequency of 125 KHz region and globally accepted ISM 2.4 GHz normal operating frequency mechanism which has reached the international advanced level.

The smart labels are mounted on the cabinet surface of the container, the equipment of antenna and reader are installed in the container lifting equipment and container run channel. The container transport information data can be read and written by the electronic tag reader, and the data exchange is carried out with the container management system using wired or wireless transmission through data interface, so to achieve the automatic identification and real-time management of the container. The intelligent management system is applied to the Shanghai Long Wu port and the image recognition technology is supplemented to capture the trailer number at the crossing so as to achieve the intelligent management of the port.

In 2006, the container electronic label application demonstration project between the Chinese Shanghai port and American Savannah port was designed in Shanghai port, which opens the container RFID applications between domestic and foreign ports.

(3) Container RFID Tags and Smart Container Security Devices in CIMC

In 2007, Container RFID tag application testing project from Chinese Chongqing Port to Shanghai Port was started by China International Marine Containers (Group) Co., Ltd (CIMC).The RFID tags were used in the trailers and container tags in the project and were achieved by passive ultra-high-frequency (UHF) technology. Container electronic tags are designed to meet the metal surface characteristics. A unique 64-bit ID number is engraved by laser on the outer surface and print out the bar code for each container number. It pastes on the surface of corresponding container electronic tags, so as to be effectively read and written on the container throughout the logistic chain at various environments and avoid the occasion of container literacy problems need to be resolved using multiple electronic tags. Ceramic packaging technology is used in vehicle electronic tag and the tag is affixed to the front windshield of the vehicle, the automatic identification of the vehicle information can be reached at the crossing.

The reading and writing equipment installed in quay crane, gantry crane and stacker use wireless data transmission module to complete the data communication between the reader and the background systems, RFID reader machinery can quickly identify the container number in the storage yard and inside the port.

(4) Smart Crossing in Dalian Port, Ningbo Port and Tianjin Port

The smart crossings in Dalian Port use RFID number recognition technology and image recognition technology to identify the trailers and containers number. When the trailers go into the crossing, the RFID reader extract vehicle tag information installed on the front windshield of the trailer, the image recognition capture container number and the port management information system confirms the correct trailer going into port.

The two intelligent crossing of Ningbo and Tianjin port work in a similar way, the trailers and containers information are integrated in a tag and the label is affixed to the front windshield of the trailers. Before the trailers enter the port, the binning bill and vehicle information are inputted into the computer system and radio frequency tags at loading yard and send inputted information to the port through EDI center. While the trailer vehicles enter the crossing, the information is collected by the RFID reader device, check the packing list with the information received before, allocate the yard slot by the computer, then let the trailers pass.

Seen from this point, automatic identification technology realizes the information collection and identification of port containers and trailer vehicles. So radio frequency tags are installed on trailer vehicles and containers at the same time, and information collection and identification of the container is accomplished with container tags and trailer vehicles with vehicle tags. When the containers go into the crossing of the port, the information of containers and trailer vehicles can be collected and identified. The container information can also be collected and identified when the containers are transported to the storage yard. All this will not only effectively prevent the abuse of trailer license, but also promote the common application of container RFID technology between the different ports.

Conclusion

RFID system is used in container multimodal transportation which makes containers as the flow of information carrier, merges information flow and logistics, tracks the information and logistics of container transportation real-time, eliminates the error and leakage in the process of container transportation. It improves customs clearance speed and transportation safety, reliability and improves the service level of container transportation comprehensively. In July 2010, international organization for standardization (ISO) officially released the ISO/PAS 18186: container shipping RFID tag system and this is made by the Shanghai port group led the vice President named Bao Qi-fan. This is an international public specification approved by the international organization for standardization. Now this standard doesn't formulate provisions, communication protocol, frequencies, and physical provisions. This will cause all products manufactured according ISO standards, to have no compatibility between each industries and enterprises of different needs, different backgrounds, and different application environment and there is uncertainty when implementing the standard. So the formulation and the perfection of the communication protocols and frequency details need further study and then completed standard will be applied to the container seamless transportation.

References

Coyle, Karen.(2005). Management of RFID in Libraries. *The journal of Academic Librarianship*, 31, 486-489.

E.W.T. Ngai, T.C.E. Cheng, S. Auc, et.al. (2007). Mobile commerce integrated with RFID technology in a container depot. *Decision Support Systems*, 43, 62–76.

Francesco Rizzo,Marcello Barboni,Lorenzo Faggion,et,al. (2011). Improved security for commercial container transports using an innovative active RFID system. *Journal of Network and Computer Applications*. 34, 846-852.

Goodrum. Paul M., Mclaren, Matt A., Durfee. Adam, (2006). The application of active radio frequency identification technology for tool tracking on construction jobsites. *Automation in Construction*, 15,292-302.

Harry K.H Chow, King Lun Choy, W. B. Lee, K. C. Lau. (2006). Design of an RFID case-based resource management system for warehouse operations. *Expert Systems with Applications*, 30, 561-576.

L. Frisk, J. Jarvincn, R. Ristolaincn. (2003). Chip on flex attachment with thermoplastic ACF for RFID applications. *Zoology*, 106, 169-171.

Laniel, M., & Émond, J.-P. (2010). Mapping of RFID tag readability in relation to the food content in a refrigerated sea container at 915 MHz, *Innovative Food Science and Emerging Technologies*, 06):1-4.

Thevissen, Patrick. (2006). Implantation of an RFID tag into human molars to reduce hard forensic identification labor. *Forensic Science International*, 159, 33-39.

Peer-reviewer: Binquan Wen
Affiliation: Tianjin Transportation University, China

In: Radio-Frequency Identification (RFID)
Editor: Xinxin Fan, pp. 45-52
ISBN: 978-1-63117-750-7

Chapter 2

RFID Technology for Drug Labelling in the Emergency Department: RFID to Minimise the Occurrence of Adverse Events

María Martínez Pérez**, *José Ramón Vizoso Hermida*, *Guillermo Vázquez González and Isabel Martín Herranz
Servicio de Informática, Xerencia de Xestión Integrada A Coruña, Spain

Abstract

RFID is currently a leading emergent technology which is being successfully implemented in many environments. It is essential to study its integration in the pharmaceutical environment in order to minimise the occurrence of adverse events and hence significantly increase patient safety and sustainability of the hospital system.

Keywords: RFID; medication unit doses; RFID labels

* E-mail address: maria.martinez.perez@sergas.es (Corresponding author)

Introduction

At present, RFID (Radio Frequency Identification) is present in very disparate fields, providing significant benefits in the performance of its tasks. RFID has aroused considerable public interest at different levels of the healthcare environment. In fact, the FDA has investigated its application in the labelling of certain types of medications [1]. It is therefore very interesting to integrate RFID in the pharmaceutical environment, with the aim of increasing the safety and quality of care received by the patient during their process of care. It is a challenge to implement RFID systems and integrate them into the daily activities of the hospital pharmacy, and for them not to be relegated to being mere pilot projects. Essentially, RFID enables radio frequency communication between a reader and a label containing identifying or other additional data about the object to which it is adhered or of the person wearing it. Its workings are simple: the RFID label generates a radio frequency signal with its information which an RFID reader captures and subsequently will send to the application responsible for managing the data. The main advantages of RFID over other identifier technologies, such as the bar code, can be found in the properties of the RFID labels, as they are not damaged with use as well as being able to simultaneously read, write and modify all the information available in the memory of hundreds of devices, without a direct view between the label and the reader. Not forgetting also, that RFID tags can withstand extreme temperatures of up to -21 ° C.An RFID system comprises hardware and software components. However, the device which undoubtedly acquires greatest protagonism are the labels. This is because their design is adapted to the characteristics of the objects or people that the system wishes to control. The successful implementation of RFID requires an extensive study to understand its functioning and the configurable parameters of its components. The aim of this paper is to present the RFID labels currently available on the market that fulfil the requirements to enable the labelling of the majority of unit doses of administered pharmaceutical submissions at the emergency department of Hospital A Coruña.

Methods

The development of this work can be divided into three main phases: the first consists in carrying out a comprehensive analysis of RFID technology. To do this, internet documents have been consulted and over 150 companies who work with this technology both nationally and internationally have been contacted. The goal is to obtain sufficient knowledge to make a successful decision regarding the selection of RFID devices.

The second phase is appointed to determine the requirements (size, shape, material, etc.) of the labels to be adhered to the unit doses of medication. In order to define these features, an analysis of 80% of the most consumed drugs over a period of three months, (randomly selected), consisting of September, October and November 2009, was performed at the Emergency Department of Hospital A Coruña.

The resulting list has been revised by the Pharmacy Department, which included those drugs considered most representatives because of the material shape of their packaging.

Finally, the third stage aims to determine the feasibility of using RFID labels in clinical practice. These were adhered to the selected medications and different readings were carried out, combining the possible values for each of its configurable parameters (reading distance, insulation and frequencies). This process has identified the strengths and weaknesses of all models a priori that are feasible for labelling medication unit doses.

RFID labels [2, 3, 4, 5, 6, 7] have many features available that must be determined by the engineers who design the RFID system to be implemented. The following summarises the more important operating parameters of the labels comprising an RFID system.

Configurable Parameters of the RFID Labels

Power Source

There are different types of labels: *passive*, which use power from the reader for their transmissions, and *active*, which dispose of their own battery and thus have their own power for transmitting. The *semi-active* labels are a special type of passive label because they have a battery but do not use it to transmit, but to power the internal circuitry of its microchip. Certain characteristics of the label as, for example, cost, size, life span or the coverage, are determined by the power source type to which it belongs.

Communication Protocol

There are different protocols to establish the communication between an RFID labels and a reader [8, 9, 10, 11, 12]:

Duplex: the label transmits its information upon receiving the signal from the reader (and throughout the duration of the signal).

Taking into account the order in which the communication is established, this protocol can be sub-divided into two types:

Half duplex: the label and reader transmit alternately.

Full duplex: communication between the reader and the label is performed simultaneously. In this case, transmission of the label is carried out at a different frequency to that used by the reader.

Sequential: the label sends its data when the reader field turns off at regular intervals. This protocol is often used with active tags because as they dispose of their own power supply, they do not make use of the power sent by the RFID reader.

Data Storage Capacity

This parameter determines the amount of data that can be recorded in the memory of the tags.

Ability to Write

When manufactured, passive RFID labels are recorded with a unique and unmodifiable (UID) indentifier, setting it apart from any other passive label available on the market. Given the possibility of modifying the data available in the memory of the labels, these components can be classified into three main groups:

Reading alone: the label data is recorded at the time of manufacture and cannot be re-written again.

Written Once and multiple readings: label data can be written only once, subsequent to its time of manufacture. *Reading/writing*: label data can be written on multiple occasions, subsequent to its time of manufacture.

Speed and Reading Time of Label Data

RFID technology is characterised and differentiated from other technologies by its storage capacity and ability to modify the data recorded in the memory of its labels. Thus the speed of the RFID system when reading the data available on its labels is a factor which directly affects its output.

In general terms, the higher the operating frequency, the faster the data transfer. This concept is closely related to the availability of bandwidth of the frequency ranges used for communication. It is important to remember that it is not advisable to select very high bandwidths since as this grows, so will the level of noise received. The reading time of the available data on a label directly depends on the speed and the amount of data that has to be transmitted. The speed and time of data reading are the parameters most influenced by the RFID system operating frequency.

Coverage

Coverage of an RFID component is defined as the maximum distance at which communication can be established with it (for reading or writing data in its memory, modifying its settings, etc..). Coverage depends on factors other than the operating frequency of the system such as the power provided by the reader antenna and that available on the label, the surrounding conditions, the direction of the label at the time of reading, etc.

Requirements of the Selected RFID Label

Having analysed the characteristics of the unit doses of the selected drugs, the following requirements have been identified:

Power Source

Passive label: due to the high cost of active labels and large dimensions of these components.

Communication Protocol

Duplex communication protocol: determined by the use of passive labels.

Size

Measure less than 3 * 3 cm as there are very small sized unit doses.

Material

Reading of the label needs to be performed without interference even if attached to materials like metal because *blister*[1] packs of tablets contain aluminium, liquids

[1]Cardboard holder with a sheet of transparent plastic forming different cavities used as packaging of small manufactured goods.

in glass like syrups, or liquids in plastic like serum. This requires some type of insulation such as epoxy, foam or paper to eliminate the interference of RFID with these materials.

Reading Capability

Label with the ability to write and store in order to record the code, name, batch and expiry date of the medication to be identified (128 characters – 16 from the label UID).

Temperature

Label that can withstand low temperatures since some products are stored in a refrigerator at temperatures of below zero (up to 21 ° C) so the RFID device and the adhesive used must withstand such temperatures.

Coverage

Low frequency RFID system because the readings will be performed at close ranges, (regardless of whether there is a direct contact between the label and the reader).

Some possible passive tags for labelling unit doses of drugs are:

Epoxy encapsulated labels, mu-chip (the smallest design found), a conventional RFID passive tag without insulation, as the epoxy encapsulation substantially raises the price of the device (2.76 € per label). In contrast, however, the cost of the foam is far lower.

Other options would be specific labels for labelling glass objects, and another passive round, one whose smallest size is 28 mm. or a label specifically designed for pharmaceutical applications and the possibility of entering passive tags into the medicine bottle.

Samples of passive labels as detailed above were acquired and medications on the list of drugs supervised by the Pharmacy Service were labelled.

After having carried out tests, the label seen in Figure 1.has proven feasible in 90% of the readings taken of the materials described above.

This device operates at the HF frequency and the model is called Philips I-Code SLI which complies with ISO 15693 and its measurements are of 15 x 15 mm.

It should be noted that it is necessary to include a foam insulation for medications whose packaging contains some type of metal, thereby eliminating interference that could arise in the readings, due to the problems in RFID with certain materials.

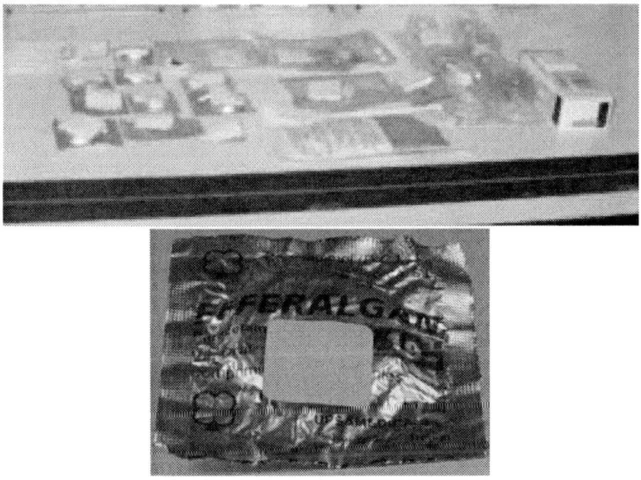

Figure 1. Passive RFID label selected.

Discussion

The cost of passive tags is quite high when compared to the price of the most frequently consumed drugs at an Emergency Department but can be justified for the labelling of expensive medications or those that are very dangerous for the patient.

This is because in addition to the cost of the label, which will decrease over time as with the growth of any technology, the work of adhering each of the labels and of physically recording in its memory the necessary data in each case must be taken into account (certain insulators can prevent the automatic printing of the label on a printer). Therefore, the ideal situation is that the drug packaging is labelled with RFID at the laboratory at the time of manufacture.

Another option is to assess the labelling of batches of medications or of the container of medicines to be administered to the patient, this would allow a notable reduction in the costs of implementing RFID and would facilitate the use of this technology in hospital clinical practice

Acknowledgments

This project was funded by the Instituto de Salud Carlos III, Spain (project PI10/02442 andproject PI07/90351).

References

Abdallah, M.A., Lei, Z.M., Li, X., Greenwold N., Nakajima, S.T., Jauniaux, E., Rao, C.V., 2004. *Human Fetalnongonadal tissues contain human chorionicgonadotropin/luteinizing hormone receptor* http://upcommons.upc.edu/pfc/bitstream/2099.1/7340/1/M%c3%a1rketing%20RFID.pdf (pp. 119 y 145).

Bolotnyy, L. y Robins, G. (2006). Generalized yoking-proofs for a group of RFID tags. *En Proceedings of the 3rd Annual International Conference on Mobile and Ubiquitous Systems*, pp. 1–4.

Huang, H. H. y Ku, C. Y. (2008). A RFID grouping proof protocol for medication safety of inpatient. *Journal of Medical Systems*, 33(6), pp. 467-474.

Juels, A. (2004). Minimalist cryptography for low-cost RFID tags. *Proceedings of the Fourth Conference on Security in Communication Networks* (SCN '04), pp. 149–164.

Juels, A. (2004). Yoking proof for RFID Tags. *En Proceedings of the Second IEEE Annual Conference on Pervasive Computing and Communications Workshops* (PERCOMW'04), pp. 138–143.

Juels, A. (2009). Yoking-proofs for RFID tags. *En Proceedings of Second IEEE Annual Conference on Pervasive Computing and Communication Security* (PERCOMW '04), pp. 138–143.

Peris-Lopez, P. Hernandez-Castro, J. C. Estevez Tapiador, J. M. y Ribagorda, A. (2006). LMAP: A real lightweight mutual authentication protocol for low-cost RFID tags. Workshop on RFID and Lightweight Crypto (RFIDSec'06). Consultada el 21 de Enero de 2012, en http://events.iaik.tugraz.at/RFIDSec06/Program/

Piramuthu, S. (2007). Protocols for RFID tag/reader authentication. *Decission Support Systems*, 43(3), pp. 897–914.

Piramuthu, S. (2007). Protocols for RFID tag-reader authentication. *Decission Support Systems*, 43(3), pp. 897–914.

Saito, J. y Sakurai, K. (2005). Grouping proof for RFID tags. *En Proceedings on 19th International Conferenceon Advanced Information Networking and Applications* (AINA 2005), 2, pp. 621-624.

Selwyn, P. (2006). On existence proofs for multiple RFID Tags. *En Proceedings of IEEE International Conference on Pervasive Services* (ICPS'06), pp. 317-320.

Tu, Y.-J. y Piramuthu, S. (2007). RFID distance bounding protocols. *Proceedings of the First International EURASIP Workshop on RFID Technology* (RFID2007), pp. 67–68.

In: Radio-Frequency Identification (RFID)
Editor: Xinxin Fan, pp. 53-70
ISBN: 978-1-63117-750-7

Chapter 3

APPLICATIONS OF RFID TECHNOLOGY IN THE CONSTRUCTION INDUSTRY: AN ACADEMIC LITERATURE REVIEW

Javad Majrouhi Sardroud *
Faculty of Engineering, Islamic Azad University,
Central Tehran Branch, Tehran, Iran

Abstract

Although other industries profit from Radio Frequency Identification (RFID) technology in managerial tasks, construction projects are late adopter of this technology. Construction today is becoming increasingly complex. It is therefore under greater pressure than ever before to become more integrated.

Adopting RFID could provide real-time and accurate information to undertake efficient and effective planning as well as execution of the projects. In order to bridge the gap between researchers and practitioners, this paper aims to summarise available information related to the RFID applications in construction. Challenges which need to be addressed before we can fully benefit from RFID are highlighted.

Keywords: Automation, Construction management, RFID technology

---

* E-mail address: J.Majrouhi@gmail.com.

Introduction

Construction projects are extremely complex and often take place in an uncontrolled and dynamic environment where each project goes through several phases leading to completion. Modern construction management requires a real-time and accurate sharing of information amongst all parties involved to undertake efficient and effective planning as well as execution of the projects (Majrouhi Sardroud and Limbachiya, 2011). There are numerous important challenges facing today's construction industry that are encouraging the adoption of Information and Communication Technologies (ICT). A chronic problem in the construction industry that requires urgent attention is to improve real-time information visibility and traceability, which leads to the success of construction projects (Lu et al., 2011). Thus, the availability of accurate information at the right time and in the right format to the right person will be guaranteed to support decision making.

One of the current significant issues involved with ICT is Automated Data Collection (ADC) technologies; the use of which has grown in recent years. ADC technologies can be defined as the use of automated advanced and data storage technologies for the identification, collection, storage, transmission, and presentation of information. In order to respond to the needs of projects and to ensure performance, real-time information systems have become an important tool in management. Although ADC technologies offer new opportunities to establish effective management tools, and other industries profit from technology in the managerial tasks, real-life projects in construction are late adopters of ADC technologies.

Construction projects need to be more intelligent and integrated in adopting these technologies as it is recognised as a promising solution to overcome the complexity of construction projects. In addition, according to recent research, the need and necessity for the real-time controlling of ongoing construction projects is increasingly growing (Majrouhi Sardroud, 2013). These technologies also have the potential to revolutionise productivity in the construction industry by using effective communication and ubiquitous data exchange (Majrouhi Sardroud et al., 2012a). As far as the availability of advanced technologies is concerned, a number of these, which are appropriate for construction, are becoming more effectual. Their accuracy, reliability, and integrity are improving, while their costs continue to decline (Deng et al., 2001).

One of the more recent advancements in these technologies is Radio Frequency Identification (RFID) technology. RFID has been identified as one of the ten greatest contributing technologies of the 21st century (Motamedi and Hammad, 2009) because of its potential benefits such as ready availability, ease to handling, and affordability. RFID is a wireless sensor technology, based on the detection of electromagnetic signals and radio frequencies, which are used to capture and transmit data from or to a tag (Domdouzis et al., 2007). In other words, RFID is a method of remotely storing and retrieving data by utilising radio frequency in identifying, tracking, and detecting various objects that simplify identification-- (Ergen and Güven, 2009).

Academic research into RFID in construction has proliferated significantly over the last few years as shown in Figure 1.

However, to continue with the advancement of knowledge in this area, it is important to understand the current status of RFID research in this area to date and to examine the contemporary trends in the research domain. In order to bridge the gap between researchers and practitioners, the prime objective of this paper is to comprehensively review and summarise available literature on RFID applications in construction. This is an attempt to facilitate the creation and accumulation of knowledge in this subject area.

RFID Technology: An Overview

A RIFD system consists of tags (transponder) with an antenna, a reader (transceiver) with an antenna, and a host terminal (Figure 2).

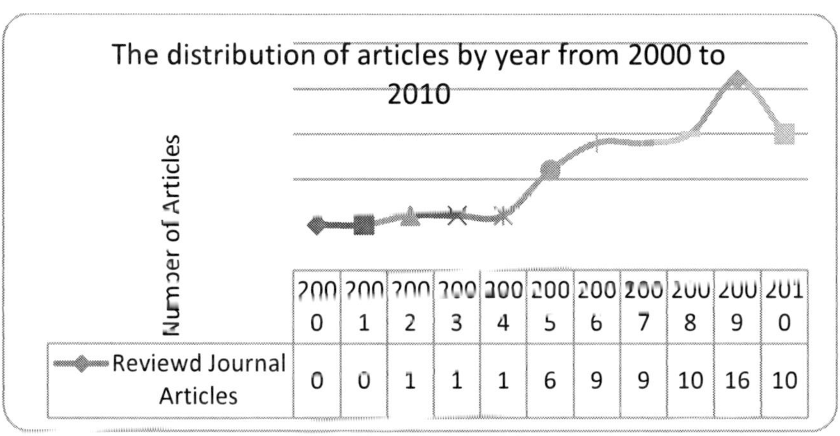

Figure 1. Distribution of articles on RFID in construction by year of publication.

Figure 2. RFID system components.

An RFID tag is a portable memory device located on a chip that is encapsulated in a protective shell and can be attached to the object. Tags are able to store dynamic information about the objects. Tags consist of a small integrated circuit chip coupled with an antenna to enable them to receive and respond to radio frequency queries from a reader. Tags can be categorised as read-only (RO), write once, read many (WORM), and read-write (RW) in which the volume capacity of their built-in memories varies from a few bits to thousands of bits. RFID tags can be classified into active tags (battery powered) and passive tags, which powered solely by the magnetic field emanated from the reader.

Reading and writing ranges are depend on the operation frequency (low, high, ultra high, and microwave). Low frequency systems generally operate at 124 KHz, 125 KHz or 135 KHz. High frequency systems operates at 13.56 MHz and ultra high frequency (UHF) use a band anywhere from 400 MHz to 960 MHz (Erabuild, 2006). Tags operating at ultra high frequency (UHF) typically have longer reading ranges than tags operating at other frequencies. Similarly, active tags have typically longer reading ranges than passive tags. Tags also vary by the amount of information they can hold, life expectancy, recycle ability, attachment method, usability, and cost. Communication distance between RFID tags and readers may decrease significantly due to interferences by steel objects and moisture in the vicinity, which is commonplace in a construction site. In contrast with passive tags that have an unlimited lifetime, active tags have internal battery source and therefore have shorter lifetime of approximately three to ten years (Jaselskis and El-Misalami, 2003).

The reader, combined with an external antenna, reads and writes data from or to a tag via radio frequency and transfers data to a host computer. In passive RFID, the RFID reader acts as a transmitter and receiver, which transmits an electromagnetic field that ''wakes-up'' the tag and provides the power required for the tag to operate (Wang et al., 2007). According to and Finkenzeller (Finkenzeller, 2010) the reader can be configured either as a handheld or a fixed mount device. The host and software system is an all-encompassing term for the hardware and software component that is separate from the RFID hardware (i.e., reader and tag); the system is composed of the following four main components: Edge interface or system, Middleware, Enterprise back-end interface, and Enterprise back end (Jaselskis et al., 1995).

Advantages and Benefits of RFID Technology

According to Majrouhi Sardroud and Limbachiya (Majrouhi Sardroud, 2012), the significant advantage of all types of RFID systems is the non-contact, non-line-of-sight nature of the technology for reading and writing. Unlike a bar code, RFID tags can be read almost instantaneously through other materials and they can be read through plastic, cardboard, and wood. Theoretically, this allows a pallet of mixed products, containing individual RFID tags, and equipped with an RFID reader to read all the tags within the palletised load without having to physically move any of the materials or open any cases. The RFID tags can store data and also be read in challenging circumstances at remarkable speeds, which in most cases responding in less than 100 milliseconds.

In any RFID system, multiple tags can be read in one time, which leads to an increase in the efficiency of data processing. In interactive applications such as work-in-process or maintenance tracking, the read and write capability of an active RFID system is also a significant advantage. RFID equipment damage occurs much less frequently than is the case with magnetic strips or barcodes. RFID tags are reusable and less susceptible to damage and can be read through a variety of substances such as ice, snow, paint, fog, crusted grime, and other visually and environmentally challenging conditions, where barcodes or other optically read technologies would be useless. The combination of all above mentioned advantages will provide quick access to a wealth of information, eliminate human errors, and reduce labour which lead to reduce project activity times and to save project costs.

RFID tags are more durable and suitable for a construction site environment in comparison with Barcodes, which are easily peeled off and may be illegible when they become dirty (Majrouhi Sardroud et al., 2012b). Developments in RFID technology continue to yield larger memory capacities, wider reading ranges, and faster processing.

RFID Usage in Different Industries and Its Applications in Construction

There are various applications of RFID technology in different industries; manufacturing; food; defence; pharmaceutical; transport; retail and healthcare are just some of the sectors where RFID has already been extensively applied (Ngai et al., 2008).

In the construction industry, RFID has been used in some efforts to improve tracking, delivery, receipt and location of materials and components, such as fabricated pipes, structural steel members, engineered-to-order components, and interior decorating materials in lay down yards and under shipping (Majrouhi Sardroud, 2012). In addition, RFID technology has been used to develop a traceability system between paving operation and the asphalt mixing plant and implemented to improve the management of health and safety in construction. Furthermore, RFID has been used in simulation-based framework for modelling information flow processes in highway project and to facilitate the accurate 3D data of underground infrastructure such as buried assets (Majrouhi Sardroud et al., 2012b).

Aim and Scope of the Study

The aim of this paper is to understand the state of RFID applications in the construction industry by examining the published articles. This provides insights surrounding automation for construction practitioners and researchers on the major historical trends in, implications of, and future directions for RFID in the construction industry. The scope of this investigation is limited to 10 years.

This period is deemed to be representative of the RFID applications in construction literature. It is believed that journals are the resources, which are most commonly used to acquire information and release new approaches. Thus, conference papers, doctoral and master theses, textbooks, news reports, and unpublished working papers are excluded.

To accomplish the research aims, this investigation is based on the examination of various journals, all of which are related to automation and RFID applications in the construction industry. A total of 63 papers were identified and reviewed, which were published in 15 journals from various disciplines. For ease of discussion, the review has been classified into 4 major categories, namely: construction materials and components management; construction equipments, tools, and machinery management; construction progress measurement and performance control; and general applicability of RFID in construction.

Practical Issues Associated with the Use of RFID in Construction

Jaselskis et al. (1995) is one of the earliest research papers found on RFID in construction in which they concluded that: *"RFID technology is a promising technology for the construction industry that can be integrated into systems that can track materials, identify vehicles, and assist with cost controls"*.

As previously stated, all identified articles classified into 4 major categories. The distribution of articles on RFID in construction by key application areas is shown in Figure 3. It shows that the majority of the articles are related to RFID applications in construction materials and components management (38 out of 63 or 60% of the total). To describe Figure 3 in more detail, the next subsections classify all published papers with their related tasks in construction.

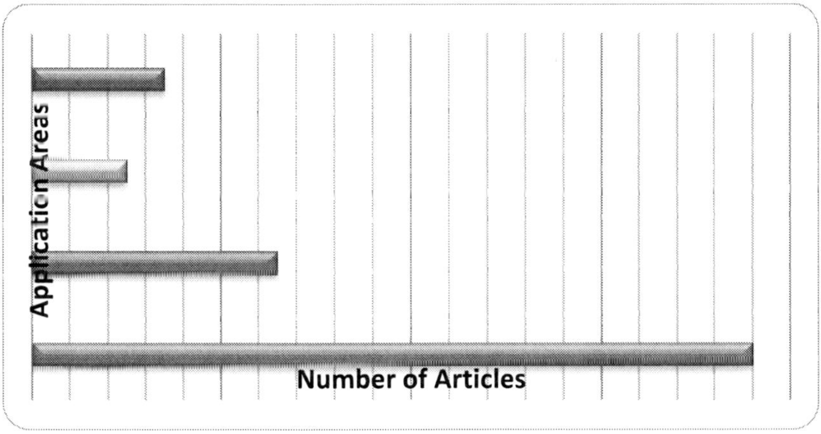

Figure 3. Distribution of articles on RFID applications in construction by application areas.

Construction Materials and Components Management

Materials management is a distinct management system, which can make significant contributions to the cost effectiveness of construction projects. Materials play an essential role in construction to complete projects within the project deadlines and budget. Some authors have developed RFID based methods to automate the task of construction materials management (Caron et al., 2007, Grau et al., 2009, Jang and Skibniewski, 2008, Jang and Skibniewski, 2009, Majrouhi Sardroud and Limbachiya, 2010a, Navon and Berkovich, 2005, Navon and Berkovich, 2006, Razavi and Haas, 2010, Song et al., 2006a, Song et al., 2007, Sørensen et al., 2010, Wan and Kumaraswamy, 2009). RFID has been used in some efforts to improve tracking, delivery, receipt, and locating of materials and components such as fabricated pipe (Song et al., 2006b), precast concrete components (Ergen et al., 2007c), structural steel members (Torrent and Caldas, 2009), curtain wall installation (Yu et al., 2007), engineered-to-order components (Ergen et al., 2007b), interior decorating materials (Tzeng et al., 2008) in lay down yards and under shipping in construction. In addition, RFID has been used in some research to improve quality management performance (Kang and Gandhi, 2010, Moon and Yang, 2010, Wang, 2008).

Furthermore, RFID has been used in some research efforts to improve the process of capturing quantity of work data at a construction site in terms of accuracy and completeness to eliminate secondary tasks due to missing or inaccurate data such as parts and packets unified architecture and linking virtual models with physical objects (Umetani et al., 2006, Yagi et al., 2005), simulation-based framework for modelling information flow processes in highway project (Akinci et al., 2006), ubiquitous system for context-specific information delivery (Behzadan et al., 2008), mobilised information frame in production management system (Yin et al., 2009), lifecycle management of facilities and building components (Motamedi and Hammad, 2009, Cheng et al., 2008), and indoor localisation (Pradhan et al., 2009).

Locating tagged items effectively in construction supply chain and logistics can potentially facilitate great increases in productivity where each project goes through several phases leading to completion (Majrouhi Sardroud and Limbachiya, 2010b). In order to obtain real-time information and information sharing among the involved participants of the construction supply chain some research efforts have been made to develop a system by using RFID technology (Wang et al., 2007, Omar and Ballal, 2009).

The findings from some research demonstrated that locating buried assets by applying RFID technology has a great potential for facilitating the accurate 3D data of underground infrastructure (Dziadak et al., 2008, Dziadak et al., 2009). This could lead to a decrease in the risk during excavation by construction and renovation equipment. In addition, RFID has been used in some research for improve facility management during their life service (Kiziltas et al., 2008, Lindkvist and Elmualim, 2009, Taneja et al., 2010) and to enhance building and facility maintenance management (Ergen et al., 2007a, Ko, 2009).

Construction Equipments, Tools and Machinery Management

As the scale of construction projects increases, efficient use of machinery and equipments becomes much more important for completion of projects within defined budget and deadline (Majrouhi Sardroud and Limbachiya, 2010c). Furthermore, tool availability is one of the universal factors and has significant potential to improve the productivity of construction workers. In response to this situation, researchers have developed automatic system based on RFID technology to improve the efficiency of tracking tools (Goodrum et al., 2006), tracking the movement of construction equipment (Lee et al., 2006, Lu et al., 2009, Navon and Shpatnitsky, 2005, Sacks et al., 2005, Shehab, 2009, Tam et al., 2007) and vehicles (Lu et al., 2007) on and off construction sites.

In addition, RFID technology has been used to develop a traceability system between paving operation and the asphalt mixing plant (Peyret and Tasky, 2004) and implemented to improve the management of health and safety in construction (Chae and Yoshida, 2010, O'Brien et al., 2009, Riaz et al., 2006, Wu et al., 2010).

Construction Progress Measurement and Performance Control

The aim of project performance control is to identify discrepancies between actual and desired performance of a construction project. Thus, real-time control of ongoing construction projects based on high quality data is vital to identify these deviations. Performance in construction is measured in terms of project performance indicators, such as schedule, and cost. Some research efforts developed RFID based project models to enhance performance control (Navon and Goldschmidt, 2002, Navon, 2005) and to improve progress measurement and management in construction (Chin et al., 2008, El-Omari and Moselhi, 2009, Hu, 2008).

General Applicability of RFID in Construction

RFID technology has been used in some other applications such as general applicability of technology in construction (Jaselskis and El-Misalami, 2003, Menzel et al., 2008, Shen et al., 2010, Sørensen et al., 2009, Wing, 2006), reliable security and safety system for the construction site (Abderrahim et al., 2005), and managing construction document information (Elghamrawy and Boukamp, 2010).

Conclusion

The literature review shows great potential for RFID technology in improving construction efficiency and effectiveness, leading to the enhancement of the project management triangle (cost, time, and quality) together with safety and environment. Although some of the aforementioned approaches provide satisfactory levels of accuracy, in general there are two major drawbacks which limit their use in wide range applications such as real construction sites. Firstly, all are either dependent on pre-installed infrastructure (e.g. routers) or controlled environmental conditions. Secondly, their data transmission capability is usually limited to a small range compared to a wide area in which a large scale operation typically takes place (e.g. a construction project).

In addition, the literature review shows that a few studies evaluated the performance of RFID with field tests for the proposed applications. Other studies that tested the performance of RFID focused only on one process in one phase, and mostly related to certain segments of the entire process. Therefore, the results of the tests cannot be generalised for other processes. In most of the tests described in those studies, components or pieces of equipment were manually scanned with handheld readers; the technical feasibility of a fully automated approach was neither proposed nor tested.

Although researchers have developed their applications and systems by combining the RFID with some other technologies such as GPS, 4D CAD, ZigBee, Wireless Sensor Network, GIS, Barcode, Portal system, Dead reckoning (DR) and GPRS, this study focuses only on RFID technology. Future studies are recommended for further exploration of the above technologies which are used in conjunction with RFID technology. There are four main challenges, which need to be addressed before we can fully benefit from the nascent and ubiquitous RFID technology. These are as follows: Data on a tag or in a database? , Frequency (LF, HF, or UHF) and technology selection (passive or active)? Identification or locating, outdoor or indoor? and RFID inspection and monitoring itself?

- Data on a tag or in a database?
  There is an ongoing debate about storage of data. As stated earlier in this paper, RFID tag has an on-board data storage capacity. However, in some reviewed research, object-related information was kept in a database. In total, there are three scenarios for storing data in any RFID application. These are storing the necessary data on a tag, for example as implemented by (Song et al., 2006b, Dziadak et al., 2008, Goodrum et al., 2006), storing data in a database, implemented by (Wang et al., 2007, Ko, 2009, Chin et al., 2008), and an integrated application, combining the previous two approaches, developed by (Yin et al., 2009, Cheng et al., 2008, Hu, 2008). In the integrated approach some of the data needs to be stored on the tag, while the remaining information should be kept in a database.
- Frequency (LF, HF, or UHF) and technology selection (passive or active)?
  As stated earlier, although tags operating at UHF bands typically have longer reading ranges than LF and HF, performance reduction is observed around metals and liquids when using UHF. However, if tags are required to be embedded in steel element, the LF band should be chosen.
  There is another ongoing debate about frequency and technology selection for RFID application in construction. Some researchers have developed RFID applications by choosing passive technology in low frequency (Chin et al., 2008, Hu, 2008), high frequency (Lee et al., 2006), and ultra high frequency (Tzeng et al., 2008). In contrast, with this research, other researchers implemented RFID applications by choosing active technology in high frequency (Yin et al., 2009) and ultra high frequency (Tzeng et al., 2008, Chae and Yoshida, 2010). The combination of both passive and active approaches has been used in some research by (Ergen et al., 2007c, Ergen et al., 2007b, Wu et al., 2010). In some instances, for the same application, tagging methods were different. In some research, RFID tags have been attached inside the object or element (Wang, 2008), while other research placed the tag outside of the object (Jaselskis and El-Misalami, 2003).
- Identification or locating, outdoor or indoor?
  There are no clear messages about technology performance for choosing as a suitable technology for location tracking. In addition, there is no information as to whether technology is more appropriate for indoor or outdoor applications.
- RFID inspection and monitoring itself?

Applications of any technology in construction should be accompanied by its inspection and monitoring guidelines. It is also necessary to provide guidelines to run inspection and monitoring processes for the technology itself during the implementation phase.

As previously stated, the greatest challenge in implementing RFID applications in real-world construction projects is to bridge the gap between researchers and practitioners. Complexity of all these challenges not only necessitates the consideration of the technological features of RFID technology itself, but also the creation of a flowchart for any application and the preparation of a clear guideline for its implementation. This will help practitioners decide the right concept for the right conditions, most appropriate for their needs. In some cases, the comparison between construction and other industries may help the practitioners in deciding the right concept. Furthermore, advantages and disadvantages of any approach should be identified and taken into account. However, in all scenarios a cost-benefit analysis should also be performed to choose the right concept.

References

Abderrahim, M., Garcia, E., Diez, R. and Balaguer, C. 2005. A mechatronics security system for the construction site. *Automation in Construction,* 14, 460-466.

Akinci, B., Kiziltas, S., Ergen, E., Karaesmen, I. Z. and Keceli, F. 2006. Modeling and analyzing the impact of technology on data capture and transfer processes at construction sites: a case study. *Journal of Construction Engineering and Management,* 132, 1148-1157.

Behzadan, A. H., Aziz, Z., Anumba, C. J. and Kamat, V. R. 2008. Ubiquitous location tracking for context-specific information delivery on construction sites. *Automation in Construction,* 17, 737-748.

Caron, F., Razavi, S. N., Song, J., Vanheeghe, P., Duflos, E., Caldas, C. H. and Haas, C. T. 2007. Locating sensor nodes on construction projects. *Autonomous Robots,* 22, 255-263.

Chae, S. and Yoshida, T. 2010. Application of RFID technology to prevention of collision accident with heavy equipment. *Automation in Construction,* 19, 368-374.

Cheng, M. Y., Tsai, H. C., Lien, L. C. and Kuo, C. H. 2008. GIS-based restoration system for historic timber buildings using RFID technology. *Journal of Civil Engineering and Management,* 14, 227–234.

Chin, S., Yoon, S., Choi, C. and Cho, C. 2008. RFID+4D CAD for progress management of structural steel works in high-rise buildings. *Journal of Computing in Civil Engineering,* 22, 74-89.

Deng, Z. M., Li, H., Tam, C. M., Shen, Q. P. and Love, P. E. D. 2001. An application of the Internet-based project management system. *Automation in Construction,* 10, 239-246.

Domdouzis, K., Kumar, B. and Anumba, C. 2007. Radio Frequency Identification (RFID) applications: A brief introduction. *Advanced Engineering Informatics,* 21, 350–355.

Dziadak, K., Kumar, B. and Sommerville, J. 2009. Model for the 3D location of buried assets based on RFID technology. *Journal of Computing in Civil Engineering,* 23, 148–159.

Dziadak, K., Sommerville, J. and Kumar, B. 2008. RFID based 3D buried assets location system. 13. Available: http://www.itcon.org/cgi-bin/works/Show?2008_11.

El-Omari, S. and Moselhi, O. 2009. Data acquisition from construction sites for tracking purposes. *Engineering, Construction and Architectural Management,* 16, 490-503.

Elghamrawy, T. and Boukamp, F. 2010. Managing construction information using RFID-based semantic contexts. *Automation in Construction,* 19, 1056-1066.

Erabuild 2006. RFID in construction; *Review of the current state of Radio Frequency Identification (RFID) technology, its use and potential future use in construction*. EU: National Agency for Enterprise and Construction, Tekes, Formas and DTI.

Ergen, E., Akinci, B., East, B. and Kirby, J. 2007a. Tracking components and maintenance history within a facility utilizing radio frequency identification technology. *Journal of Computing in Civil Engineering,* 21, 11-20.

Ergen, E., Akinci, B. and Sacks, R. 2007b. Life-cycle data management of engineered-to-order components using radio frequency identification. *Advanced Engineering Informatics,* 21, 356-366.

Ergen, E., Akinci, B. and Sacks, R. 2007c. Tracking and locating components in a precast storage yard utilizing radio frequency identification technology and GPS. *Automation in Construction,* 16, 354-367.

Ergen, E. and Guven, G. Assessing the need for storing data on Radio Frequency Identification (RFID) Tags. *26th International CIB W078 Conference - Managing IT in Construction*, 2009.

Finkenzeller, K. 2010. *RFID Handbook: Fundamentals and Applications in Contactless Smart Cards and Identification,* Wiltshire, UK, John Wiley and Sons, Ltd.

Goodrum, P. M., Mclaren, M. A. and Durfee, A. 2006. The application of active radio frequency identification technology for tool tracking on construction job sites. *Automation in Construction,* 15, 292-302.

Grau, D., Caldas, C. H., Haas, C. T., Goodrum, P. M. and Gong, J. 2009. Assessing the impact of materials tracking technologies on construction craft productivity. *Automation in Construction,* 18, 903-911.

Hu, W. 2008. Integration of Radio Frequency Identification and 4D CAD in Construction Management. *Tsinghua Science and Technology,* 13, 151-157.

Jang, W. S. and Skibniewski, M. J. 2008. A wireless network system for automated tracking of construction materials on project sites. *Journal of Civil Engineering and Management,* 14, 11-19.

Jang, W. S. and Skibniewski, M. J. 2009. Cost-benefit analysis of embedded sensor system for construction materials tracking. *Journal of Construction Engineering and Management,* 135, 378–386.

Jaselskis, E. J., Anderson, M. R., Jahren, C. T., Rodriguez, Y. and Njos, S. 1995. Radio frequency identification applications in construction industry. *Journal of Construction Engineering and Management,* 121, 189-196.

Jaselskis, E. J. and El-Misalami, T. 2003. Implementing radio frequency identification in the construction process. *Journal of Construction Engineering and Management,* 129, 680-688.

Kang, J. and Gandhi, J. 2010. Readability test of RFID temperature sensor embedded in fresh concrete. *Journal of Civil Engineering and Management,* 16, 412-417.

Kiziltas, S., Akinci, B., Ergen, E., Tang, P. and Gordon, C. 2008. *Technological assessment and process implications of field data capture technologies for construction and facility / infrastructure management*. 13. Available: http://www.itcon.org/cgi-bin/works/Show?2008_10.

Ko, C. H. 2009. RFID-based building maintenance system. *Automation in Construction,* 18, 275–284.

Lee, U. K., Kang, K. I., Kim, G. H. and Cho, H. H. 2006. Improving tower crane productivity using wireless technology. *Computer-Aided Civil and Infrastructure Engineering,* 21, 594-604.

Lindkvist, C. and Elmualim, A. 2009. Pervasive technologies for workspace management. *Journal of Facilities Management,* 7, 98-110.

Lu, M., Chen, W., Shen, X., Lam, H. C. and Liu, J. 2007. Positioning and tracking construction vehicles in highly dense urban areas and building construction sites. *Automation in Construction,* 16, 647-656.

Lu, M., Shen, X. and Chen, W. 2009. Automated collection of mixer truck operations data in highly dense urban areas. *Journal of Construction Engineering and Management,* 135, 17-23.

Lu, W., Huang, G. Q. and Li, H. 2011. Scenarios for applying RFID technology in construction project management. *Automation in Construction,* 20, 101-106.

Majrouhi Sardroud, J. 2012. Influence of RFID technology on automated management of construction materials and components. *Scientia Iranica, Transaction A: Civil Engineering,* 19, 381-392.

Majrouhi Sardroud, J. 2013. Quality inspection and management of construction projects using RFID technology. *Iranian Journal of Science and Technology, Transactions of Civil Engineering,* 37, 469-478.

Majrouhi Sardroud, J., Bromhead, E. N. andShojaee, A. A.Implementing RFID Technology in Quality Inspection and Management of Dam Construction. *The First International and Third National Conference On Dams and Hydropower*, 2012a.

Majrouhi Sardroud, J., Hosseyni, S., Bromhead, E. N. and Riazi, M. 2012b. Automated landslide monitoring and warning using RFID technology. *International Journal of Safety and Security Engineering,* 2, 118-130.

Majrouhi Sardroud, J. and Limbachiya, M. C. 2010a. Effective information delivery at construction phase with integrated application of RFID, GPS and GSM technology. *Lecture Notes in Engineering and Computer Science,* 2183, 425-431.

Majrouhi Sardroud, J. and Limbachiya, M. C. Improving construction supply chain management with integrated application of RFID technology and portal system. *8th International Conference on Logistics Research* (RIRL 2010), 2010b.

Majrouhi Sardroud, J. and Limbachiya, M. C. 2010c. Utilization of advanced data storage technology to conduct construction industry on clear environment. *Journal of World Academy of Science, Engineering and Technology,* 66, 808-813.

Majrouhi Sardroud, J. and Limbachiya, M. C. 2011. Towards linking islands of information within construction projects utilizing RF technologies. In: GELMAN, L. and AO, S. I. (eds.) *Electrical Engineering and Applied Computing.* Springer.

Menzel, K., Zixiang, C. and Allan, L. 2008. Potentials for Radio Frequency Identification in AEC/FM. *Tsinghua Science and Technology,* 13, 329-335.

Moon, S. and Yang, B. 2010. Effective monitoring of the concrete pouring operation in an RFID-based environment. *Journal of Computing in Civil Engineering,* 24, 108-116.

Motamedi, A. and Hammad, A. 2009. *Lifecycle management of facilities components using radio frequency identification and building information model.* 14. Available: http://www.itcon.org/cgi-bin/works/Show?2009_18.

Navon, R. 2005. Automated project performance control of construction projects. *Automation in Construction,* 14, 467-476.

Navon, R. and Berkovich, O. 2005. Development and on-site evaluation of an automated materials management and control model. *Journal of Construction Engineering and Management,* 131, 1328-1336.

Navon, R. and Berkovich, O. 2006. An automated model for materials management and control. *Construction Management and Economics,* 24, 635-646.

Navon, R. and Goldschmidt, E. 2002. Monitoring labor inputs: automated-data-collection model and enabling technologies. *Automation in Construction,* 12, 185-199.

Navon, R. and Shpatnitsky, Y. 2005. A model for automated monitoring of road construction. *Construction Management and Economics,* 23, 941-951.

Ngai, E. W. T., Moon, K. K. L., Riggins, F. J. and Yi, C. Y. 2008. RFID research: An academic literature review (1995–2005) and future research directions. *International Journal of Production Economics,* 112, 510–520.

O'brien, W. J., Julien, C., Kabadayi, S., Luo, X. and Hammer, J. 2009. *An architecture for decision support in Ad Hoc sensor networks.* 14. Available: http://www.itcon.org/cgi-bin/works/Show?2009_21.

Omar, B. and Ballal, T. 2009. *Intelligent wireless web services: Context-aware computing in construction-logistics supply chain.* 14. Available: http://www.itcon.org/cgi-bin/works/Show?2009_20.

Peyret, F. and Tasky, R. 2004. A traceability system between plant and work site for asphalt pavements. *Computer-Aided Civil and Infrastructure Engineering,* 19, 54-63.

Pradhan, A., Ergen, E. and Akinci, B. 2009. Technological assessment of radio frequency identification technology for indoor localization. *Journal of Computing in Civil Engineering,* 23, 230–238.

Razavi, S. N. and Haas, C. T. 2010. Multisensor data fusion for on-site materials tracking in construction. *Automation in Construction,* 19, 1037-1046.

Riaz, Z., Edwards, D. J. and Thorpe, A. 2006. SightSafety: A hybrid information and communication technology system for reducing vehicle/pedestrian collisions. *Automation in Construction,* 15, 719-728.

Sacks, R., Navon, R., Brodetskaia, I. and Shapira, A. 2005. Feasibility of automated monitoring of lifting equipment in support of project control. *Journal of Construction Engineering and Management,* 131, 604-614.

Shehab, T. 2009. Automated measurement of earthwork volume in construction projects. *Construction Innovation: Information, Process, Management,* 9, 341-358.

Shen, W., Hao, Q., Mak, H., Neelamkavil, J., Xie, H., Dickinson, J., Thomas, R., Pardasani, A. and Xue, H. 2010. Systems integration and collaboration in architecture, engineering, construction, and facilities management: A review. *Advanced Engineering Informatics,* 24, 196-207.

Song, J., Haas, C. T. and Caldas, C. H. 2006a. Tracking the location of materials on construction job sites. *Journal of Construction Engineering and Management,* 132, 911-918.

Song, J., Haas, C. T. and Caldas, C. H. 2007. A proximity-based method for locating RFID tagged objects. *Advanced Engineering Informatics,* 21, 367-376.

Song, J., Haas, C. T., Caldas, C. H., Ergen, E. and Akinci, B. 2006b. Automating the task of tracking the delivery and receipt of fabricated pipe spools in industrial projects. *Automation in Construction,* 15, 166-177.

Sørensen, K. B., Christiansson, P. and Svidt, K. 2009. *Prototype development of anict system to support construction management based on virtual models and RFID.* 14. Available: http://www.itcon.org/cgi-bin/works/Show?2009_19.

Sørensen, K. B., Christiansson, P. and Svidt, K. 2010. Ontologies to support RFID-based link between virtual models and construction components. *Computer-Aided Civil and Infrastructure Engineering,* 25, 285-302.

Tam, C. M., Tong, T. and Wong, B. 2007. An integrated system for earthmoving planning. *Construction Management and Economics,* 25, 1127-1137.

Taneja, S., Akinci, B., Garrett, J. H., Soibelman, L., Berges, M., Atasoy, G., Liu, X., Shahandashti, S. M., Anil, E. B., Ergen, E., Pradhan, A. and Tang, P. 2010. CEC: Sensing and Field Data Capture for Construction and Facility Operations. *Journal of Construction Engineering and Management,* doi:10.1061/(ASCE)CO.1943-7862.0000332.

Torrent, D. G. and Caldas, C. H. 2009. Methodology for automating the identification and localization of construction components on industrial projects. *Journal of Computing in Civil Engineering,* 23, 3-13.

Tzeng, C. T., Chiang, Y. C., Chiang, C. M. and Lai, C. M. 2008. Combination of radio frequency identification (RFID) and field verification tests of interior decorating materials. *Automation in Construction,* 18, 16-23.

Umetani, T., Arai, T., Mae, Y., Inoue, K. and Maeda, J. I. 2006. Construction automation based on parts and packets unification. *Automation in Construction,* 15, 777-784.

Wan, S. K. M. and Kumaraswamy, M. M. 2009. Industrial management approaches for improving material control in building services works. *Engineering, Construction and Architectural Management,* 16, 208-223.

Wang, L. C. 2008. Enhancing construction quality inspection and management using RFID technology. *Automation in Construction,* 17, 467-479.

Wang, L. C., Lin, Y. C. and Lin, P. H. 2007. Dynamic mobile RFID-based supply chain control and management system in construction. *Advanced Engineering Informatics,* 21, 377-390.

Wing, R. 2006. *RFID applications in construction and facilities management.* 11. Available: http://www.itcon.org/cgi-bin/works/Show?2006_50.

Wu, W., Yang, H., Chew, D. A. S., Yang, S. H., Gibb, A. G. F. and Li, Q. 2010. Towards an autonomous real-time tracking system of near-miss accidents on construction sites. *Automation in Construction,* 19, 134-141.

Yagi, J., Arai, E. and Arai, T. 2005. Parts and packets unification radio frequency identification application for construction. *Automation in Construction,* 14, 477-490.

Yin, S. Y. L., Tserng, H. P., Wang, J. C. and Tsai, S. C. 2009. Developing a precast production management system using RFID technology. *Automation in Construction,* 18, 677–691.

Yu, S. N., Lee, S. Y., Han, C. S., Lee, K. Y. and Lee, S. H.2007. Development of the curtain wall installation robot: Performance and efficiency tests at a construction site. *Autonomous Robots,* 22, 281–291.

In: Radio-Frequency Identification (RFID) ISBN: 978-1-63117-750-7
Editor: Xinxin Fan, pp. 71-91

Chapter 4

SOFTWARE DEFINED RADIO, GNU RADIO, RFID AND THEIR APPLICATIONS

Muhammad Khizer Kaleem*
Ericsson Canada Inc., Canada

Abstract

This chapter explains the introduction and background of software defined radio. We briefly elaborate on how SDR has evolved over the years as well as present an overview of the first publicly announced military software radio named SPEAKeasy. This chapter also describes GNU Radio (an open source software toolkit used for signal processing on a host computer) and its architecture as well as USRP (a hardware device used as a transceiver for radio signals that feeds the DSP or host embedded PC with sampled signals for processing). Later in the chapter there is a brief overview of the RFID system architecture, physical layer characteristics in different RFID standards, and the RFID communications in lower layers using USRP and WISP tag (which will justify the use of USRP (SDR) as an RFID reader). Finally, the conclusion provides experimental results of WG-7 lightweight stream cipher for RFID encryption and authentication.

Keywords: software radio, software defined radio, SDR, SR, USRP, GNU Radio

*E-mail address: muhammad.khizer.kaleem@ericsson.com

1. Introduction

Radio is a science of communicating over a distance by converting sounds or signals to electromagnetic waves and radiating these through space in a radio frequency band in a range of about 10kHZ to 300GHz [15].

With the evolution of communications technology and the rapid change of circuits from analog to digital, it became possible to implement most of the radio functions using software in modern radio systems - leading to 'software defined radio'. The term software defined radio (SDR) or software radio (SR) was first coined by Joseph Mitola in 1991. The latest definition was recently coordinated between the IEEE and the SDR Forum as follows:

A software defined radio is radio in which some or all of the Physical Layer Functions are software-defined.

An extensive definition given by Joseph Mitola is:

A software radio is a radio whose channel modulation waveforms are defined in software. That is, waveforms are generated as sampled digital signals, converted from digital to analog via a wideband DAC and then possibly upconverted from IF to RF. The receiver, similarly, employs a wideband Analog to Digital Converter (ADC) that captures all of the channels of the software radio node. The receiver then extracts, downconverts and demodulates the channel waveform using software on a general purpose processor. Software radios employ a combination of techniques that include multi-band antennas and RF conversion; wideband ADC and Digital to Analog conversion (DAC); and the implementation of IF, baseband and bitstream processing functions in general purpose programmable processors. The resulting software-defined radio (or software radio) in part extends the evolution of programmable hardware, increasing flexibility via increased programmability [9].

A software radio uses digital signal processing and software communication radio system functions (mixers, filters, amplifiers, modulators/demodulators, detectors, channel model etc.) on a targeted processor and memory to host it for real time applications.

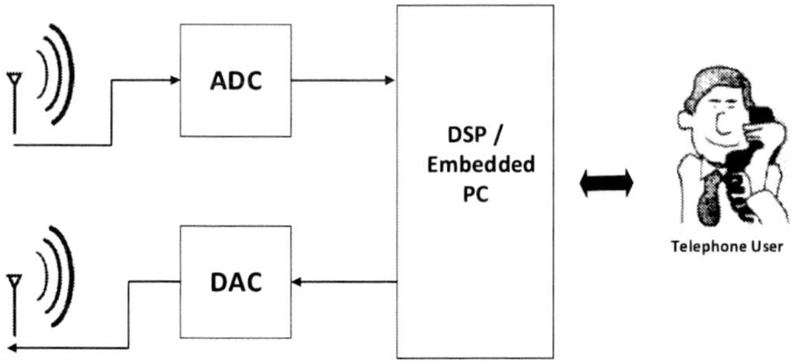

Figure 1. An ideal Software Defined Radio.

1.1. An Ideal Software Radio

An ideal software radio consists of an Antenna, Analog to Digital converter (ADC)/ Digital to Analog Converter (DAC) and embedded processor to handle all digital processing of radio signals. An ideal SR is shown in Figure 1.

In the receiver chain, an antenna receives the signals and passes them on to the ADC. Digital samples generated from the ADC are then passed through the DSP, or embedded PC, for further processing. Similarly, in the transmitter chain, digital samples are taken from the DSP or embedded PC and passed on to the DAC which feeds analog signals to the antenna for transmission. In a transceiver path, the processing block and an antenna may be shared between receiver and transmitter.

1.2. Practical Software Defined Radio

SDR presented in the previous section is simple and extremely flexible but it is not practical. The practical software defined radio includes many non-DSP hardware components such as anti aliasing filters, power handling, frequency up/down converters, etc. since current ADC/DAC and DSPs do not support direct sampling of wideband RF signals.

As technology evolves, digital signal processors (DSP) are becoming faster, conversion rates and dynamic ranges are increasing, and ADC/DAC are getting closer to the antenna. For these reasons, software defined radio is replacing the analog circuitry in conventional radio systems by taking up additional processing tasks.

2. Need for SDR

In the past, radios were used for military purposes and each one could only perform the particular task it had been designed and optimized for. If ever a need arose to use the radio for a different task, the complete hardware would have to be re-designed. This was a costly and less flexible solution which also led to compatibility issues. This problem was addressed by Velcro radios [17] who presented a solution for reducing the hardware that had to be carried around. The solution was developed on the basis of common architecture involved in all radios. They proposed to use a common motherboard and attach different peripherals (silicon chips) to it. For example, if an army commander wanted to talk to another party on a particular frequency, he would use a specific peripheral and connect it to his motherboard. He could replace the chip if he wanted to use another frequency and related protocol. This reduced the total hardware involved in conventional military radio systems, but there was limited cross-functionality when a specific chip was only compatible with a motherboard provided by the same manufacturer. An open modular hardware architecture was developed to address this problem but, communicating simultaneously with different radios were still not feasible [10].

As technology rapidly advanced, flexible software implementation of analog circuits in digital processing took over more tasks that had initially been accomplished with less flexible hardware devices. This flexibility, the software implementation of digital radio components, allows adaptive transitions to waveforms and RF frequencies not possible in previous technologies. SDR allows easy and simple enhancements to the existing system without replacing or modifying the hardware.

3. Evolution of SDR

Software Defined Radio originated in the US and European defense sectors in the late 1970s [17, 13]. By the end of the 1980s, emerging standards and incompatible protocols had made the radio technology evolution process stagnant [10]. There were compatibility and cross-functional problems in communicating on different RF frequency bands using different encoding methods and waveforms. Conventional radios were not keeping up to the rapid change in technology adoption. Similarly, the commercial use of radio systems was experiencing the same dilemma. Each manufacturer wanted to provide new features with minimized cost by keeping the legacy system architecture. This limited the radio systems to specific geographic locations around the world. The proponents of radio technology were looking for a *MultiBand MultiMode Radio* (MBMMR) system [16, 8].

Initially, MBMMR was implemented by encapsulating different RF bands and modes in a single package, that could change their configuration by changing the parameters, and were known as Velcro radios. By packing all these peripherals in the same system the size increased tremendously.[1].

To reduce the size of Velcro radios, Programmable Digital Radios (PDRs) were introduced in which programmable ROMs and ASIC processors replaced small hardware components. A prominent increase in efficiency, flexibility, and a reduction in size, weight and cost intrigued the manufacturers and researchers to shift more hardware functionality to software processing which gave rise to software defined radios. With the increase in conversion rate and dynamic range of AD/DA converters, the RF components got closer to the software processing and grew independent of analog hardware, which allowed for more flexibility in generating different waveforms without modifying the hardware.

4. SPEAKeasy

SPEAKeasy was the first publicly announced military software radio developed by the US Department of Defense. The rationale behind the SPEAKeasy program was to develop a modular, reprogrammable, and reconfigurable modem with an open architecture. The SPEAKeasy program was divided into two phases. Phase I aimed to demonstrate, through a proof-of-concept, the feasibility of software radios and the ability to inter-operate with more than 15 existing radios using MBMMR's envision; and, Phase II focused on an open architecture for the entire radio and a more formalized software structure as compared to Phase I.

4.1. SPEAKeasy Phase I

Phase I research was conducted by Hazeltine (as the major contractor), and TRW (as software subcontractor), from 1992 to 1995 [16]. This Phase was designed and implemented to provide the basic architecture for software radio and had the following capabilities:

[1]Moore's law does not apply to RF components [10]

- Interoperability between military radios
- Adoption of emerging new technology
- More flexibility without compromising the performance and power consumption
- Single wide-band operation from 2 MHz to 2 GHz
- Incorporation of information security as well as transmission security using a programmable INFOSEC processor, Cryptographic Reduced instruction Set (CYPRIS) and software implementation of the cryptographic algorithms in the multichannel SPEAKeasy INFOSEC Module (SIM)

The first practical demonstration of SPEAKeasy Phase I was carried out successfully in June of 1995. The interoperability was presented by connecting multiple existing radios using different protocols. New protocols were programmed into software and were demonstrated by two units of SPEAKeasy. The technology available at the time did not permit for a single wide band RF channel of range 2MHz-2GHz, instead the range was divided into 3 sub-bands: the first was High Frequency (*HF*), ranging from 2-30MHz; the second was Very High Frequency (*VHF*), ranging from 30-400MHz; and the last was Ultra High Frequency (*UHF*) ranging, from 400MHz-2GHz. Only the *VHF* band was implemented in Phase I for the feasibility demonstration. It was observed that software radios were practical for interoperability functions of radios and were compatible with new features of evolving technology. The general architecture of SPEAKeasy was the same as conventional radios but all the blocks were digital, apart from the front end antennas. The common hardware was reprogrammable, multi-chip module *(MCM)* was employed for basic signal processing, and CYPRIS was used for *INFOSEC* and *TRANSEC* for transmission security.

4.2. SPEAKeasy Phase II

Phase II of SPEAKeasy was a 4 year project, initiated in 1995, and Motorola was the prime contractor and designer of the wide-band RF transceiver in this phase. Its main features included an open architecture for the entire radio and a more formalized software structure as compared to Phase I. Security was incorporated in its software structure as two modules, *BLACK* (secure) and *RED* (unsecure). The primary objectives of Phase II were as follows:

- Implementation of a reconfigurable architecture
- Creation of an open and modular architecture for the entire radio system
- Support of Cross Channel Connectivity
- Emphasis on more commercial off-the-shelf (COTS) and incorporating standards
- Improved CYPRIS to handle numerous different algorithms of *INFOSEC* and *TRANSEC* simultaneous
- Reduction in radio size for improved practicality in the field

Although it was planned as a 4 year project, it was successfully demonstrated within 15 months and manufacturing was started immediately. The Land Mobile Radio (LMR) waveform was developed in less than two weeks to add to SPEAKeasy units. Furthermore, one of the units was repaired in the field using COTS [14]. As this was the first demonstration of Phase II, only 4-400MHz was exhibited but it achieved all primary objectives. The project was soon closed and no further research was carried out for SPEAKeasy.

The hardware and software of SPEAKeasy Phase I and II was later used by joint tactical radio system programs (JTRS) and the SDR forum for defining their standards.

5. GNU Radio

GNU Radio is a software development toolkit, provided with a signal processing library, for developing and experimenting with SDR concepts. GNU Radio works under GNU General Public License (GPL)[2]. The project was initiated in 2000 by Eric Blossom and is now supported by a large community of volunteers working for the development of SDR. GNU Radio is primarily supported on the Linux operating system with partial support for Mac and Windows.

[2]GNU General Public License is a free, copyleft license for software [5].

5.1. GNU Radio Architecture

Processing Blocks: Signal processing blocks take continuous data stream(s) at their input and generate data stream(s) at their output. Besides signal processing blocks, GNU Radio also provides signal sources and sinks that only produce or consume data stream(s) respectively. The GNU Radio processing library comes with a large number of signal processing blocks and modules, and it also provides flexibility for users to write their own processing blocks if desired[3]. The GNU radio framework takes care of each block's number of input samples it might require to produce the necessary output samples by using the **forecost()** method.

Data Buffers: Blocks may have different input stream rates in a flow graph but all output streams must be generated at the same rate. Blocks have input and output buffers associated with them to keep synchronization among different stream rates.

Flow Graph: The GNU Radio applications are written using flow graph. The vertices in flow graph represent the processing blocks, and edges constitute the data flow between them. Signal processing blocks are written in C++ and are connected using Python. Each block is primarily written for performing only one task to enable an open and modular software architecture. Processing blocks take infinite stream(s) of data flow at the input and provides the output stream(s) after processing. Every flow graph must have at least one source or sink. The GNU Radio uses Simplified Wrapper and Interface Generator (SWIG) to allow signal processing blocks (written in C++) to be accessed from Python.

Data Type: The data types of signal processing block streams can be of any kind. But the most common data types used in GNU Radio are:

- c - complex(float)
- f - single precision float
- s - short integer (2-byte) and
- i - integer (4-byte).

These data types also exist as vector type as some blocks take or generate vector as an input or output respectively. Data flows from one block to another in the form of bits, or samples.

[3] Further information about how to write a signal processing block can be found at [1, 2]

5.2. GNU Radio Companion (GRC)

GRC is a graphical tool used for creating signal flow graphs, by connecting different processing blocks written in GNU Radio, and generates Python script to create and control signal flow graphs using visual programming language for signal processing via GNU Radio libraries. It provides an easy and editable drag and drop flow graph environment to work with, without (changing a paradigm shift) worrying about the underlying processes. GRC was developed by Josh Blum [7].

6. Universal Software Radio Peripheral (USRP)

USRP provides a hardware interface for GNU Radio to build software defined radio using UHD (Universal Hardware Driver). UHD is a hardware host driver and API for Ettus Research products, i.e., all USRP devices that are compatible with Linux, Windows, and Mac. Users are able to use UHD driver as a standalone, or with third party applications such as GNU Radio, Simulink, OpenBTS, Iris and LabVIEW [4]. A typical USRP consists of a motherboard that supports several plug-on daughterboards of different RF bands. USRP with GNU radio provides a complete SDR platform.

6.1. Features of USRP-1

USRP-1 includes the following features:

- Four 64 MS/s 12-bit analog to digital converters and four 128 MS/s 14-bit digital to analog converters
- Four digital downconverters and two digital upconverters with programmable decimation and interpolation rates, respectively
- One High-speed USB 2.0 interface (480 Mb/s) capable of processing signals up to a bandwidth of 16 MHz
- Modular architecture supporting a wide variety of RF daughterboards
- Auxiliary analog and digital I/O supporting complex radio controls such as RSSI (Received Signal Strength Indication) and AGC (Automatic Gain Control)

- Fully coherent multi-channel systems (MIMO capability)

Wrap-up: GNU Radio, with the USRP, provides an inexpensive, flexible, and powerful supporting tool, in license free radio bands, for learning and experimenting with software defined radio concepts. Most applications can be developed using GRC without writing any code. With an increasing number of available signal blocks, GNU radio is becoming more and more flexible.

7. RFID Architecture

An RFID system consists of RFID devices (Tags), Readers (or interrogators) and a Back-end system (or database). The general RFID architecture is shown in Figure 2.

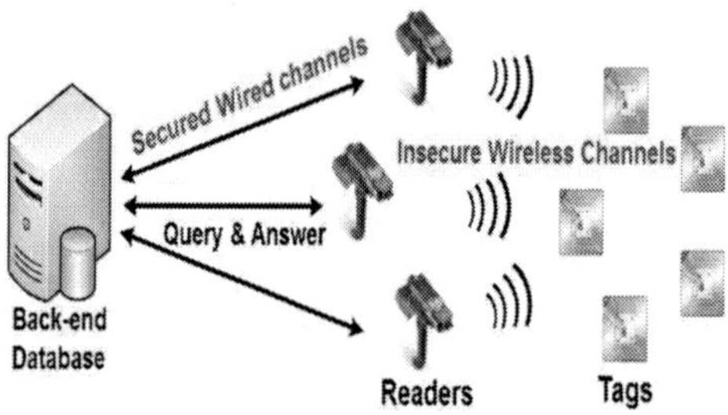

Figure 2. General Architecture of RFID System.

7.1. Tags

Tags are embedded in all objects that are to be identified in RFID systems. Modern tags contain an Integrated Circuit (IC) chip with computational and storage capabilities, and an antenna coil, or coupling element, for communications. The resources available to the tag vary depending on its power source and type. Tags can be mainly categorized into three types depending on the power source, i.e. active, passive, and semi-passive tags.

- ***Active Tags*** have an on board power source and much greater computational and storage capabilities, e.g., NFC enabled cell phones and an embedded RFID wireless sensor node.

- ***Passive Tags*** do not have an internal power source. They harvest power from the reader's signal. These tags cannot perform complex computations due to this power limitation. They harvest power when they come in the vicinity of a reader's signal and store that power, which depletes rapidly. The power used for transmitting a signal from tag to reader, is inversely proportional to the square of the distance between them.

- ***Semi-passive Tags*** lie between the active and passive tags. They use battery power for computation and harvest energy from reader's signal for communications, e.g., Automated toll payment tags

7.2. Readers

An RFID reader (or interrogator) is a transceiver that sends commands to tags and receives data from them. There are two types of readers:

1. *Portable readers*, i.e., NFC enabled cell phones acts as RFID readers. These readers have power limitations which restricts their operation range.

2. *Fixed readers* have a much greater capability for storage and processing range.

7.3. Back-end Database

A Back-end database contains private credentials for authentication of valid tags. Readers talk to database through a secured link, i.e. the secured link ensures the security of the communication between database and readers. Reader queries the tags for public IDs and then retrieves the private data corresponding to that ID from secure database. This reduces the tag cost and helps in managing the data.

7.4. Physical Layer

Digital communication system consists of three main components: signal processing unit (line coding); communication channel; and, modulator/demodulator (as shown in Figure 3). Similarly, in order for an RFID system to communicate on a physical layer it must have these three basic functional blocks. To transmit a message over an RF channel the message source must

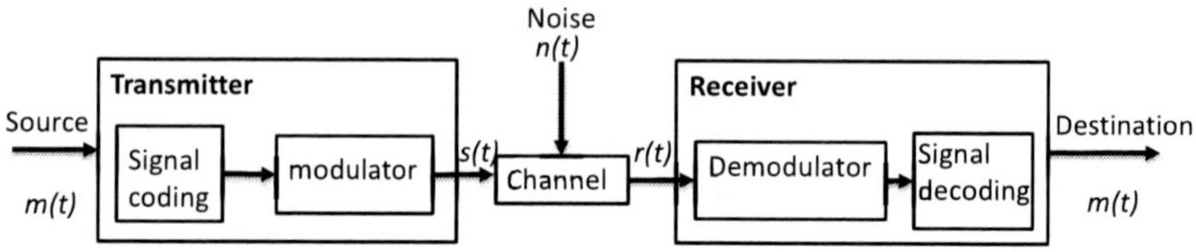

Figure 3. Signal and data flow in digital communication.

be first encoded. The encoded signal influences the electromagnetic waves of a particular carrier, depending on the type of modulation used in a system. In digital communications, the signal can be represented as a '1' or '0' in various line codes.

7.4.1. Line Coding

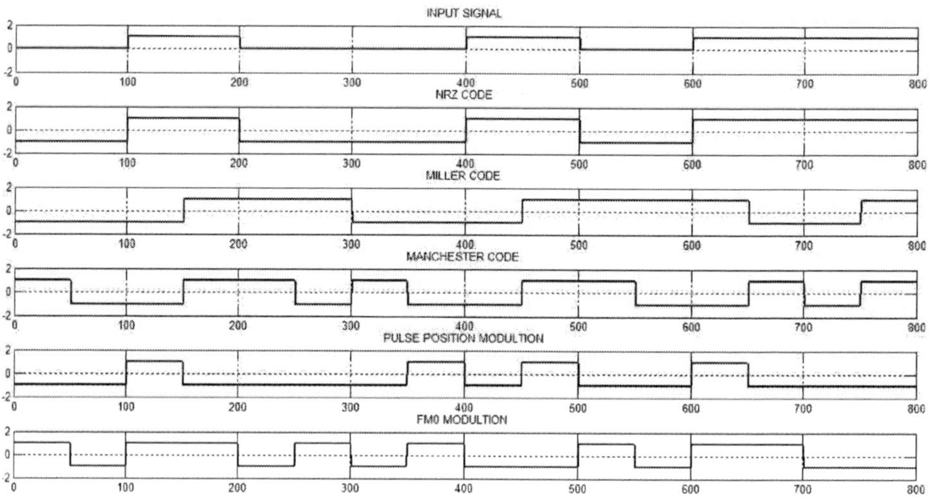

Figure 4. Signal Coding in Baseband.

Line coding is used for clock synchronization, elimination of DC components and can help in data recovery. RFID systems use one of the following schemes: Non-Return-Zero (NRZ), Manchester, Unipolar RZ, Differential Bi-Phase (DBP), Miller, Modified Miller, differential coding on Pulse Pause (PP) coding. Line coding (signal coding) used in an RFID system greatly depends on the boundary conditions. Various line coding schemes used in RFID systems are shown in Figure 4.

- ***Non-Return-to-Zero (NRZ):*** Binary bits '1' and '0' are mapped onto high signal '+1' and low signal '-1' respectively for one bit period. If NRZ encoding is used with 100% ASK the tag cannot harvest the power through the carrier. Even if 33% ASK is used with long zero run sequence, it would interrupt the power, therefore it is used with FSK and PSK modulations.

- ***Miller:*** A binary bit '1' is encoded to half bit positive/negative edge transition and bit '0' is represented by the continuance of the preceding signal level of a bit period. An initial reference is taken from the starting bit of the sequence and the following bits are then mapped onto high signal and low signal depending on the combinations of preceding and current bits. If '1' bit is followed by '1', the half bit edge transition would be converted to its inverse transition. If '0' bit is followed by '0', inverse transition occurs at the start of bit period.

  In EPC Class-1 Generation-2 UHF RFID protocol, tag to reader communication uses a modified miller sub-carrier coding [3]. A bit '0' is always encoded with low pulse at the start of a bit period and bit '1' is encoded with low pulse with a delay of half bit period. Miller modulated sub-carrier encoding uses 2, 4 or 8 cycles of a sub-carrier square wave for each bit period. The modified Miller encoding transitions are indicated by phase shifts of 180° in the sub-carrier. Each sub-carrier is modulated with ASK or PSK on another carrier.

- ***Manchester:*** A low to high and high to low transition are represented by bits '1' and '0' respectively. These transitions occur in the middle of a cycle, which helps in data clock recovery. Manchester coding techniques utilize twice the bandwidth of NRZ with exclusively no DC component.

- ***Pulse Position:*** Pulse Position Modulation (PPM) encodes m message bits in one of the 2^m time slots. PPM is suitable for reader to tag communications since the decoding circuit is extremely simple, which results in small and light weight tags.

- ***Pulse Interval Encoding:*** In Pulse Interval Encoding (PIE), also known

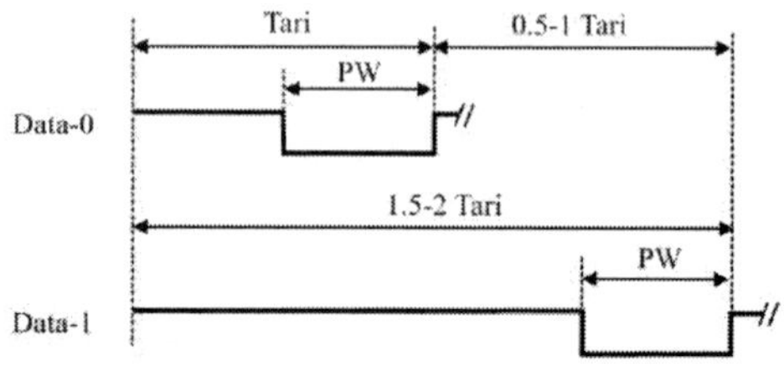

Figure 5. Baseband waveform of Pulse Interval Encoding [3].

as pulse-pause coding, a bit ‘1’ is represented by a pause of $2t$ before the next bit period and a bit ‘0’ is represented by a pause of t before the next bit period where t is a unit time duration varying from $6.35\mu s$ to $25\mu s$ also known as Tari (Type A reference interval) as shown in Figure 5. PIE is used in EPC C1 Gen2 UHF RFID protocol from reader to tag communications [3]. The advantage of using PIE from reader to tag data communication is that there is a continuous flow of power supply to the tag.

- ***FM0 Encoding:*** FM0 encoding is a bi-phase-space encoding. In FM0 encoding, a bit ‘1’ encodes to a constant signal level and a bit ‘0’ encodes to transition in the middle of a bit period. FM0 inverts the phase at the boundary of each bit period. FM0 encoding is used in tag to reader communications in EPC C1 Gen2 UHF RFID protocol [3]. Since there is a transition in each bit period in FM0, the data clock is embedded in it and can be recovered easily.

7.4.2. Modulation

Modulation is the process of sending a baseband data signal by varying the characteristics of a periodic electromagnetic wave (*carrier signal*). A sinusoidal carrier can be characterized by

$$x(t) = A_i \sin(2\pi f_c t + \phi_i) \tag{1}$$

From equation (1), the different carriers can be generated by varying the carrier's amplitude A_i, frequency f_c, and phase ϕ_i to represent the baseband data signal. In HF and UHF RFID systems, digital modulations schemes are used, i.e., *Amplitude Shift Keying* (ASK), *Frequency Shift Keying* (FSK) and

Phase Shift Keying (PSK). Examples of modulation schemes used in ISO 14443 and EPC UHF C1 Gen2 are shown in Figure 6.

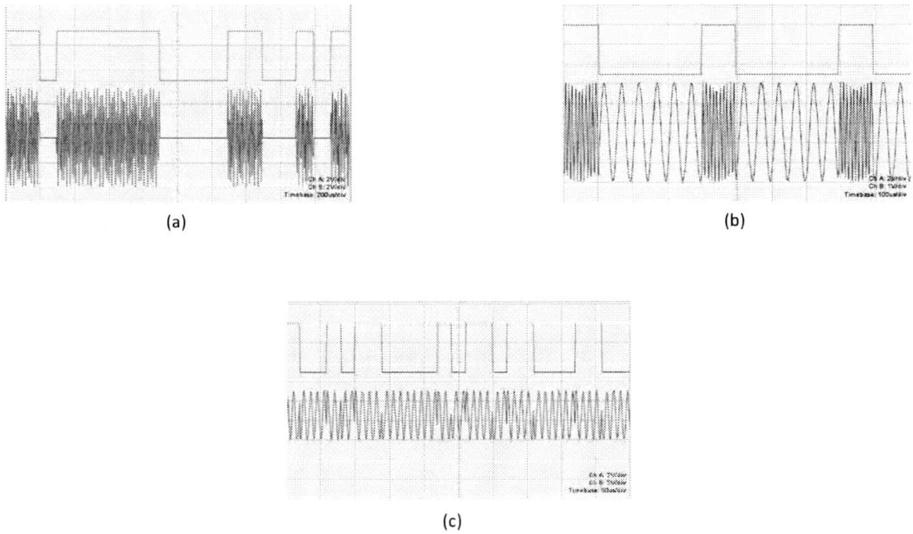

Figure 6. (a) Amplitude Shift Keying (*ASK or OOK*) (b) Frequency Shift Keying (*FSK*) (c) Phase Shift Keying (*PSK*).

In the ISO 14443 framework, type A uses 100% ASK modulation at 13.56 MHz and a sub-carrier at 847 kHz with a modified miller encoding. Type B uses BPSK modulation at same the frequency as type A with NRZ-L encoding. In UHF C1 Gen2 standard, reader to tag communications use 30% to 100% ASK (*DSB ASK, SSB ASK, PR-ASK*) at 902-928 MHz divided into channels (200kHz, 500kHz) with PIE encoding. Tag to reader communications use ASK or PSK with FM0 or Modified Miller encoding

7.5. Backscatter Communications

The basic principle on which passive RFID system work is similar to the radar principle in which the reader sends RF signal to the passive tag. The passive tag harvests power from the incident continuous wave sent by the reader to communicate back to the reader. The readers send an amplitude modulated carrier given by

$$CW(t) = A(t)\cos(2\pi f_i t + \theta) \tag{2}$$

where $A(t)$ is the amplitude of the carrier wave, f_i is the carrier frequency $\in \{f_1, ..., f_k\}$ and θ is a phase of a carrier signal.

This CW is rectified at the tag and supplies power to the tag. The tag decodes the reader's command if any – using an envelope detector. After transmitting the command, the reader keeps $A(t)$ at high amplitude while waiting for the tag response. In EPC C1 Gen2 [3], the tag could reply on the same CW with ASK or PSK modulations using FM0 or Miller sub-carrier encoding schemes with clock recovery information embedded in it. The tag backscatters the encoded responses, mixed with CW, by changing the reflection coefficient of the antenna. The received signal is detected using a matched filter and the backscattered signal's amplitude is compared to a threshold.

8. USRP for Emulating an RFID Reader

Michael and David (in [11, 12]) designed a low cost RFID reader and monitor based on the USRP for communicating and experimenting with C1 Gen2 RFID system. The GNU Radio based software architecture uses built in blocks provided by the toolkit‘s signal processing library. USRP based RFID readers need custom blocks along with standard blocks to fully implement the Gen2 protocol. The block diagram of USRP based reader architecture is shown in Figure 7.

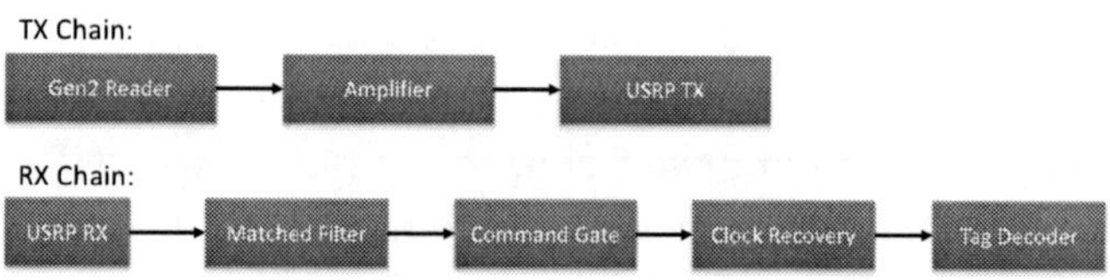

Figure 7. Block diagram of USRP reader architecture.

Transmitter (TX) Chain: The *Gen2 Reader* block implements the Finite State Machine (FSM) of the Gen2 and encodes the generated reader's commands to On/Off keying encoding for transmission. Commands are then fed to a configurable *Amplifier* and sent to the USRP hardware using UHD which provides hardware interface between USRP and a host PC. Finally, the reader commands are upconvented, converted to analog and then transmitted through an antenna.

Receiver (RX) Chain: The response from a tag is captured by *USRP RX* which downconverts the received signal and feeds it to the flowgraph. The received samples are detected using *Matched Filter* and the tag transmission is compared to a threshold. This block is configured to maximize the signal-to-noise ratio (SNR) of the tag's response. The filtered signal is then sent to the *Command Gate*. If the signal contains the tag's response the *Command Gate* passes the incoming signal to the *Clock Recovery* which re-samples the tag's response and outputs one sample per subcarrier cycle. The *Tag Decoder* block detects the preamble and demodulates the tag's ASK modulated message. The decoded bits, along with a message, are sent to *Gen2 Reader* block which changes the reader's FSM and sends out the next command.

8.1. Gen2 Authentication between a WISP Tag and a USRP Reader

In EPC C1 Gen2 standard, the QUERY is carried out in the following fashion:

Reader Tag

$$
\begin{array}{|c|}
------------Query------------>\\
<------------RN16-------------\\
------------ACK(RN16)----------->\\
<-------------EPC--------------\\
--------------NAK------------->
\end{array}
$$

In the general RFID model, the reader continuously inventories tags in its vicinity by broadcasting a $Query$ message. An inventory round starts with the reader transmitting a $Query$ command which specifies the number of slots in the frame. After receiving the $Query$, the tags randomly choose a slot in which to reply, in order to avoid collisions with other tags, and responds with a 16-bit $RN16$ random number. Upon receiving the $RN16$, the reader $ACKs$ with $RN16$ and the tag replies with a 96-bit EPC identifier. Tags that collide in a given slot are replied to with $NAKs$ and subsequently respond in the next $Query$ round.

Luo *et al.* (in [18]) proposed a lightweight stream cipher (WG-7) for RFID encryption and authentication. The proposed mutual authentication protocol works as follow:

Reader | Tag

– – – – – – – – – – – – – – $Query$ – – – – – – – – – – – – – >
< – – – – – – – – – – – – – –$RN16\_T$ – – – – – – – – – – – – – –
– – – – – – – – – – – – – $ACK(RN16\_R)$ – – – – – – – – – – – >
Compute the authentication code
– – – – – – – – – – – – – – $Query$ – – – – – – – – – – – – – >
< – – – – – – – – – – – – – –$RN16\_T$ – – – – – – – – – – – – – –
– – – – – – – – – – – – – –$ACK(0001)$ – – – – – – – – – – – – >
< – – – – – – – – – – – – –$EPC(AUTH\_CODE)$ – – – – – – – –
– – – – – – – – – – – – – – NAK – – – – – – – – – – – – – >
verify

Due to the stringent responding timing required by EPC C1 Gen2, the authentication protocol use two $Query$ sessions to complete one authentication (since the computation of the authentication code, even with the lightweight stream cipher (WG-7), takes a while).

The Mutual Authentication Protocol QUERY is Carried Out as Follows:

The reader sends out a $Query$ command to wake up the tag. The tag responses with a 16-bit random number $(RN16\_T)$. The reader sends an ACK command to the tag, where the payload is another 16-bit random number $(RN16\_R)$. The tag starts to compute the authentication code, i.e., $E(RN16\_R||RN16\_T, K)$, where K is the pre-shared secret key between the reader and the tag, and E is the WG-7 stream cipher. After a while, the reader sends another $Query$. If the tag finishes the computation of the authentication code it responds with the same $RN16\_T$. Otherwise, it gives no response and waits for the next $Query$ round. The reader then sends out another ACK command with a special payload "0001" to inform the tag to send back the authentication code produced. The tag sends back the 96-bit authentication code. The reader clears up the session by sending a NAK command and checks if the received authentication code is valid. If so, the tag authenticates itself to the reader.

8.2. Implementation Results

Figure 8 shows the experimental results of authentication with the WG-7 stream cipher for RFID encryption and authentication. Figure 8 part (a) shows the

```
Started Cycle
R=>T   Query
R<=T   RN16_T: 02D7
R=>T   RN16_R: 9C71
R=>T   Query
R=>T   Query
R<=T   RN16_T: 02D7
R=>T   RN16_R: 0001
R=>T   EPC: 9412A98E8A2786FBD5B9DB62
(verification succeeded!)
```

(a) Valid Tag

```
Started Cycle
R=>T   Query
R<=T   RN16_T: 02D7
R=>T   RN16_R: F441
R=>T   Query
R<=T   RN16_T: 02D7
R=>T   RN16_R: 0001
R=>T   EPC: 666666010000000000410002
(verification failed!)
```

(b) Invalid Tag

Figure 8. Gen2 Authentication between a WISP Tag and a USRP Reader.

communication between RFID reader and C1 Gen2 validated tag, and part (b) shows communication with an invalid tag. These experiments are conducted using USRP1 equipped with two RFX900 daughterboards, one for transmission and one for reception. The experimental setup for demonstrating Gen2 authentication is shown in Figure 9.

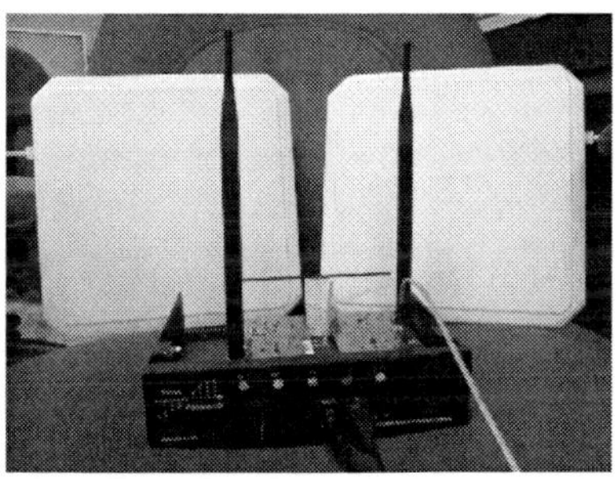

Figure 9. Gen2 authentication experimental setup.

The C1 Gen2 tag has been realized using an Intel WISP tag [6], which implements all the Gen2 commands. The firmware of WISP tag is modified to implement WG-7 stream cipher encryption and authentication. The WISP tag decoded message is shown in Figure 10.

Acknowledgment

The author would like to thank Qi Chai for providing the code used in experimentation with WG-7 stream cipher based encryption and authentication on USRP reader and Intel WISP tag. Besides, the author would like to express his appreciation to Jeffrey Benn for his constructive criticism.

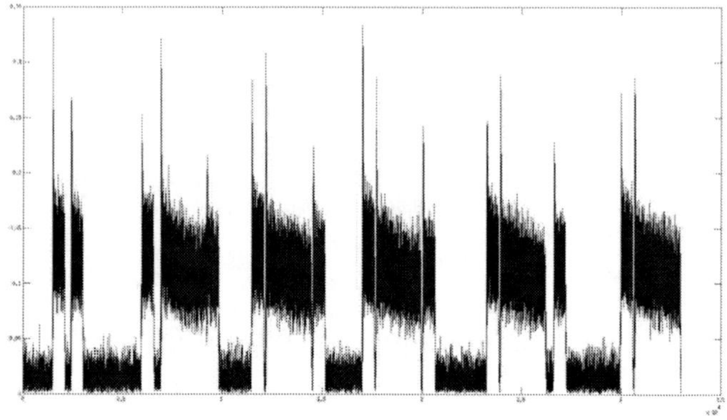

Figure 10. WISP tag's decoded message at receiver's side.

References

[1] Dawei Shen. *Tutorial 10: Writing A Signal Processing Block for GNU Radio - Part I*, June 11, 2005. Available http://www.snowymtn.ca/GNURadio/GNURAdioDoc-10.pdf

[2] Dawei Shen. *Tutorial 11: Writing A Signal Processing Block for GNU Radio - Part II*, June 15, 2005. Available http://www.snowymtn.ca/GNURadio/GNURAdioDoc-11.pdf

[3] EPC Global, *Class-1 Generation 2 specification for RFID air protocol standard v1.2.0*, http://www.epcglobalinc.org/, 2008.

[4] Ettus Research LLC, *Universal Software Radio Peripheral*. https://ettus-apps.sourcerepo.com/redmine/ettus/projects/uhd/wiki, 2013.

[5] *GNU GENERAL PUBLIC LICENSE*, Version 3. http://www.gnu.org/licenses/gpl.html, 29 June 2007.

[6] *Intel WISP*, http://wisp.wikispaces.com, 2013.

[7] Josh Blum. *GNU Radio Companion.* http://www.joshknows.com/grc, 2013.

[8] J. Mitola, "The software radio architecture", *IEEE Commun. Mag.*, vol. 33, pp.26 -38 1995.

[9] J. Mitola, *"What is a Software Radio?"* [Online] Available: http://web.archive.org/web/20050315234159/; http://ourworld.compuserve.com/homepages/jmitola/whatisas.htm 2013.

[10] J. Reed, *"Software Radio: A Modern Approach to Radio Engineering"*, Prentice Hall (2002).

[11] M. Buettner and D. Wetherall. A Gen 2 RFID monitor based on the USRP. *ACM SIGCOMM Computer Commun. Rev.*, vol. 40, no. 3, pp.41-47, 2010.

[12] M. Buettner and D. Wetherall. A software radio-based UHF RFID reader for PHY/MAC experimentation. *Proc. IEEE Int. Conf. RFID*, pp. 134-141, 2011.

[13] M. Nguyen. Software Radio (R) Evolution and Its Application to Aeronautical Mobile Communications. *Integrated CNS Conference* May 19-22, 2003.

[14] P.G. Cook and W. Bonser. Architectural overview of the SPEAKeasy system. *IEEE J. Select. Areas Commun.*, vol. 17, no. 4, pp.650 -661, 1999.

[15] Rudolf F. Graf, *"Modern Dictionary of Electronics (7th edition)"*, pp.615-616 1999.

[16] R. J. Lackey and D. W. Upmal, "Speakeasy: The military software radio", *IEEE Commun. Mag.*, vol. 33, pp.56 -61 1995.

[17] S. Gultchev, K. Moessner, D. Thilakawardana, T. Dodgson and Rahim Tafazolli. Evaluation of Software Defined Radio Technology. *Centre of Communications System Research, University of Surrey*, February 2006.

[18] Y. Luo, Q. Chai, G. Gong, X. Lai. A lightweight stream cipher WG-7 for RFID encryption and authentication. *IEEE Globecom 2010 proceedings*, pp. 1-6, 2010.

In: Radio-Frequency Identification (RFID) ISBN: 978-1-63117-750-7
Editor: Xinxin Fan, pp. 93-118

Chapter 5

ASPECTS OF RFID SECURITIES

Fei Huo*
University of Waterloo, Canada

Abstract

RFID technology has gained tremendous popularity in the recent years. The tiny, inexpensive RFID tags can be easily attached to objects for seamless identification. However, one glaring weakness of RFID tags, especially the passive RFID tags are their lack of capabilities for implementing strong crypto primitives for security purposes. Without the proper security mechanisms, the adversary could easily gain access to the tag. The secrecy and the privacy of the system are thus compromised. This is especially troublesome in the area where the RFID tag contains sensitive information such as a patient's medical record. In this chapter, we intend to give an overview of different aspects of RFID system securities, this include different attacks and countermeasures in the literature. More specifically, we first give a brief overview on the RFID system which introduces different classes of RFID systems as well as the layered design of the RFID system. Then we present different attacks corresponding to different layers along with the potential countermeasures. Finally, some future perspectives on the development of RFID security are discussed.

PACS: 05.45-a, 52.35.Mw, 96.50.Fm

Keywords: RFID, RFID security

*E-mail address: fhuo@uwaterloo.ca

AMS Subject Classification: 53D, 37C, 65P

1. Introduction

Radio frequency identification (RFID) is a promising technology which can perform the automated and unique identification of objects. Its ability to contactlessly identify distant objects as far as tens of meters away without the requirement of line of sight (LOS) is a very attractive property over some of the existing technologies such as barcode. Specifically, there are two distinct advantages of RIFD over traditional barcode technology [11]: First, RFID provides the unique identification of each object. This is because each RFID tag has its own unique identification (UID) number that distinguishes itself from all other tags. Second, the scanning of barcode would require LOS contact between the reader and the object. Careful positioning of the tag is required. Moreover, the reading range is also very limited. In the contrast, RFID readers are able to read tags from meters away without the LOS requirement. Thus RFID systems would make inventory controls much more convenient and efficient. Since its invention, RFID has found wide range of applications. This includes passport, driver's license, building access control and supply chain management just to name a few. However, RFID technology is not without its weaknesses. One of the most growing concern is its vulnerabilities against security and privacy attacks.

A typical RFID system is consisted of three components. One or multiple RFID readers, one or multiple RFID tags and a back-end database as shown in Figure 1 .

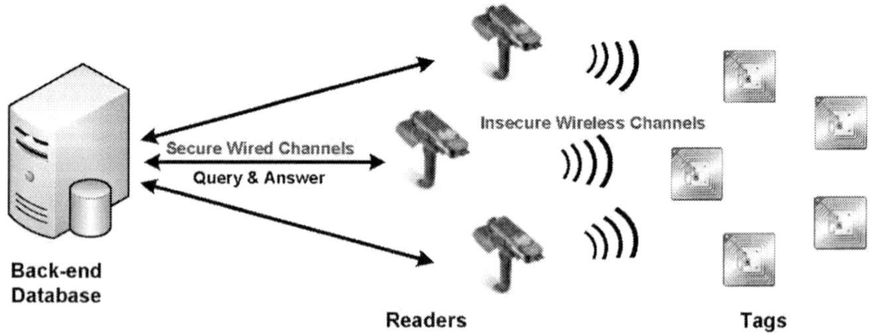

Figure 1. A Typical RFID System.

Generally, multiple readers can simultaneously communicate with different tags in a RFID system. For the simplest case, let's consider the communication between one reader and one tag. When the reader tries to communicate with a tag, it first sends a query to the tag, the tag replies with its response. The response may includes its own unique UID. Upon receiving the response, the reader transmits the collected information to the connected back-end database for further processing and verifications. The communication link is also established. Generally, RFID tags can operate in three frequencies. 124 kHz - 135 kHz low frequency tags which has a nominal read range up to half of a meter. 13.56 MHz high frequency tags which has a nominal read range of up to one meter. 860 MHz - 930 MHz ultra high frequency (UHF) tags which can read up to ten meters.

There are different RFID standards specifying the operating frequencies, modulation schemes, coding schemes and communication protocols. ISO 18000 is a multi-frequency standard which incorporates LF, HF and UHF bands. ISO 14443 specifies mostly for proximity HF RFID devices such as contactless cards used for building access control. ISO 15693 is another HF standard which has a higher reading range. Near field communication (NFC) or ISO 18092 standard is designed to be compatible with ISO 14443 and ISO 15693. This enables the NFC device to either act as a reader or as a tag. The most dominant standard for UHF RFID tags is electronic product code (EPC) Class 1 Gen 2 standard. These tags are expected to be widely adopted to replace barcode for inventory control. The organization which oversees the development of this type of tags is EPCglobal Inc. [6].

Based on the power source that drives communications between the reader and the tag, RFID tags can be classified into the following three classes [10]:

- *Active*: Tags have their own battery source and can initiate the communication.
- *Semi-Passive*: Tags have onboard battery that drives the chip's circuitry. However, power for communication is still derived from the reader's electromagnetic field.
- *Passive*: Tags have no built-in power source. The tag needs to harvest power from the reader's electromagnetic field to facilitate communications.

Active and semi-passive tags all have on-board batteries, which provide them with adequate computational capability. Passive RFID tags on the other hand are small and inexpensive, but they can only harvest power from the reader, their memory is generally also very limited. The first two classes of tags are powerful, implementations of secure cryptographic primitives on these tags are possible. It is the last class of RFID tags that requires special attentions.

1.1. Security Measures in RFID

In order for a RFID system to be considered secure, two criteria must be fulfilled. They are the reader security and the tag security [11].

Reader security: Reader security concerns with well-behaved tags disclosing unwanted information to mis-behaved readers. RFID systems are designed such that any readers within the reading range of any tags are allowed to interrogate them without notifying their owners. It is this convenience feature of RFID technology that has also brought potential security flaws that can be exploited by the adversaries. Any person carrying an RFID tag is effectively broadcasting its UID to all nearby readers. Without the privacy protection mechanisms, the information stored in this tag could be disclosed to or exploited by the mis-behaved readers without the owner's notifications. It is especially worrisome if these tags contain sensitive information such as a patient's medical or health record.

Tag security: Tag security deals with the well-behaved readers trying to obtain information from mis-behaved tags. A mis-behaved tag can impersonate the legitimate tag. The well-behaved readers may be fooled by the mis-behaved tag into believing it is communicating with a legitimate tag. The misbehaved tag could send inaccurate information or even viruses to the reader. In the later case, the entire RFID network could be at the great risk.

Cost is a major concern for a wide deployment of RFID technologies into the real world applications. Passive RFID are generally chosen over more expensive active and semi-passive tags for replacing barcodes due to the reduced cost. However, as discussed earlier, this type of tags are memory and computational constrained, standard public key crypto for encryption and hash functions for authentication cannot be implemented. Thus they are most vulnerable to various attacks. Protecting the privacy and authenticity of passive RFID tags proves to be a challenging task.

All attacks which we will discuss later on are aimed at breaking the confidentiality and authenticity protection mechanisms of the RFID system, or to completely disable the communication of the intended reader and tag.

1.2. Outline

The organization of this chapter is as follows. In Section 2, we show the design flaw on the authentication protocol of the well known EPCglobal Class 1 Gen 2 standard. In Section 3, we survey the different attacks on the RFID system. We also present the countermeasures in the literature to combat these attacks. A brief conclusion and some remarks are given in Section 4.

2. Overview of Security of EPCglobal Class 1 Gen 2 UHF Tags

In this section, we will show the design vulnerability of authentication protocol in the EPCglobal Class 1 Gen 2 standard. This attack can be conducted with relative ease. The impact of this attack could potentially be devastating as the attacker could gain access to the tag's password.

As mentioned earlier, the EPCglobal Class 1 Gen 2 standard is the dominant standard for UHF tags. It is expected to be widely adopted to replace the barcode for the inventory control among other applications. The EPCglobal Class 1 Gen 2 UHF RFID standard specifies each tag to have four memory banks [6]. They are Reserved, ECP, TID and User. Reserved memory bank is used to store 32-bit Access Password and 32-bit "KILL" password. EPC memory stores the EPC number. TID memory contains class-identifier values. It may also contain tag and vendor specific data such as tag's serial number. User memory is optional which offers user specific data storage. For detailed descriptions on the EPCglobal Class 1 Gen 2 UHF RFID standard, the reader is referred to [6].

Authentication is needed in order to prove one's identity. In the EPCglobal Class 1 Gen 2 standard, one way authentication, from the tag to the reader is implemented. The protocol is illustrated in Figure 2. The notations used in the figure are provided in Table 1.

In this authentication protocol, the reader first sends a command to the tag requiring a 16-bit random number, upon receiving the random number sent by the tag, the reader sends $CCPwd_M$ which is just the bitwise exclusive or (XOR) values of the tag's 16 most significant bits of access password with the 16-bit random number. The tag verifies the $CCPwd_M$ by performing the XOR operations again to remove the 16-bit random number and verifies with its own pass-

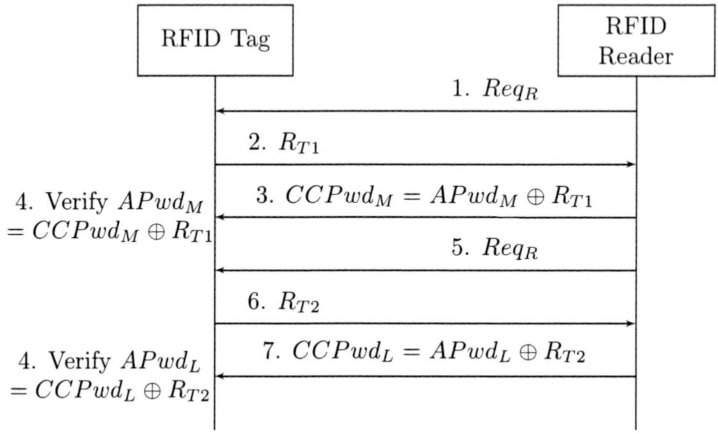

Figure 2. EPCglobal Ckass 1 Gen 2 RFID Authentication Protocol.

Table 1. Symbol Notations

Notations	Descriptions
Req_R	Command requesting 16 bit random number
R_{Tx}	16 bit random numbers generated by the tag
$APwd$	Tag's 32 bit access password
$APwd_M$	16 most significant bits of $APwd$
$APwd_L$	16 least significant bits of $APwd$

word. The entire procedure is performed again with a different 16-bit random number and the 16 least significant bits of the access password. The authentication succeeds if the verification in both rounds by the tag is successful.

One can immediately observe this authentication protocol is not secure at all! Other than the fact that the length of the APwd is only 32 bits, which can be broken quite easily using the exhaustive search attack, there exists one serious flaw with this protocol. R_{T1}, R_{T2} are un-encrypted random challenges numbers which is sent in the open channel. If the adversary intercepts R_{T1} and $CCPwd_M$, he can recover $APwd_M$. By the same reasoning, once the adversary intercepts R_{T2} and subsequent $CCPwd_L$, he can then recover $APwd_L$.

Consequently, the adversary gains knowledge to tag's all 32-bit access password. The privacy of the tag is compromised. Afterwards, the adversary could performs all kinds of attacks such as tag impersonation, modifying the tag's data, etc. This vulnerability of this protocol could put the entire EPCglobal Class 1 Gen 2 UHF RFID system at the great risk.

3. RFID Attacks and Countermeasures

In this section, we will discuss some of the existing attacks on the RFID systems. Attacks described in this section are by no means an exhaustive list of all attacks against RFID. For a more thorough description of attacks on the security and privacy of RFID systems, readers are referred to the two following survey papers [11, 22].

Attacks can take place at different layers of a RFID system. Figure 3 categorize RFID attacks at the different layers[22]. They are physical layer, net-

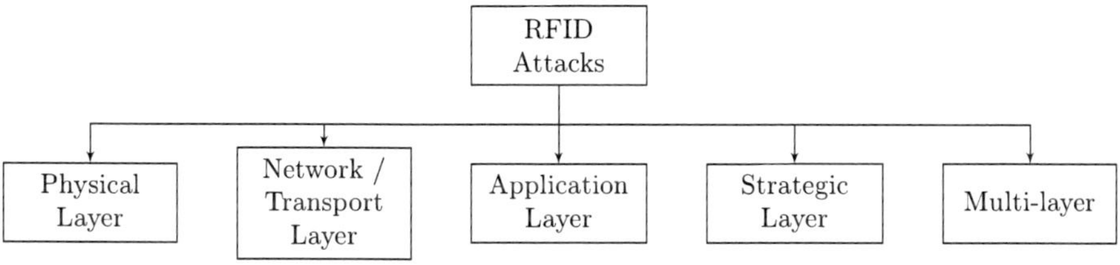

Figure 3. Categorization of RFID Attacks.

work/transport layer, application layer and strategic layer. Physical layer attacks are performed on the radio frequency (RF) signals transmitted between the reader and the tag. Network/transport layer primarily deals the communications between the reader and the tag. It also considers the exploitation of vulnerabilities of the RFID protocols. The aforementioned attack on the EPCglobal Class 1 Gen 2 authentication protocol belongs to the network/transport layer attack. Application layer attacks include exploiting the vulnerabilities of the commercial software and applications. Strategic layers include attacks related to logistical factors, they are out of the scope of this chapter. Finally, multi-layer attacks exploit the vulnerabilities from multiple layers and then form a best attack strategy.

This section will be organized as follows. We will first show the attacks in each layer, then we will present potential countermeasures in the literatures to prevent these attacks.

3.1. Physical Layer Attacks

The attack in this layer by the adversary mainly uses the wireless nature and the lack of physical security of RFID communication systems. Physical layer attacks can range from less threatening passive eavesdropping attack to more serious destruction of tags attack. Figure 4 provide some of well known physical layer attacks.

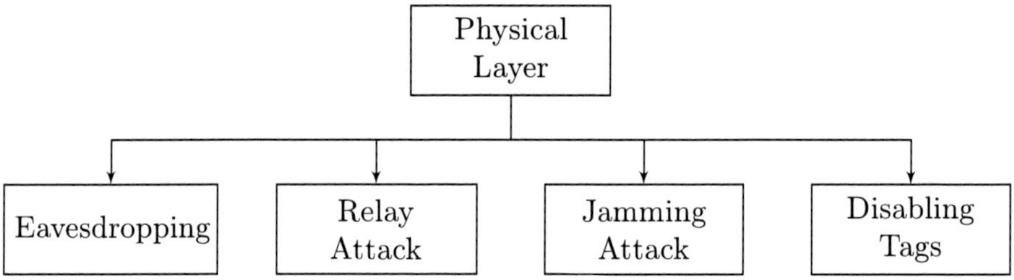

Figure 4. Physical Layer Attacks.

Eavesdropping: The adversary simply tunes his receiver to the frequency used between the tag and reader, he can capture all the communications between the reader and the tag. He may then try to recover useful information from the intercepted messages.

Relay Attack: A more serious version of eavesdropping attack is the relay attack. In the eavesdropping attack, the adversary simply listens to the ongoing communication in the channel. However, in the relay attack, the adversary intercepts the signal and then pass it on with or without modifications to the tag. The ephemeral connection is relayed from the legitimate reader to the tag through the adversary. Consequently, the legitimate receiver and tag still think they are communicating directly with each together. Francis *et. al.* have demonstrated a practical implementation of the relay attack on the NFC-enabled mobile devices [7]. Relay attacks are further categorized into *mafia fraud* and *terrorist fraud* [2].

- *Mafia fraud*: An adversary acts as the man in the middle in relaying the information between the two legitimate parities.

- *Terrorist fraud*: This is an extension of the mafia fraud in which the adversary is helped by a legitimate but dishonest tag. The legitimate but dishonest tag and the adversary collaborate together into trying to deceive the reader that it is within the close proximity to him when in fact the tag is outside the reading range of the reader.

Jamming Attack: Unlike the two previous attacks, the adversary's goal in the active jamming is no longer to capture the communications in order to gain information about the reader and tag, but instead he wants to prevent the communication between the two parties. He can achieve this goal by creating a strong noise signal at the same frequency utilized by the reader and the tag. Information received by the reader or the tag are corrupted by this strong noise signal. Decoding them back to meaningful data become very difficult if not impossible.

Disabling Tags: Disabling of tags can be achieved in a number of different ways. This include physical destruction of the tag by improper handling and purposely damaging the tag, the intentional or unintentional use of "KILL" command against the tag. These methods would permanently disable the tag. Moreover, RFID tags are sensitive to static electricity, it can be permanently damaged or temporarily rendered inoperable in a strong electromagnetic field.

3.2. Physical Layer Attack Countermeasures

In this subsection, we will present some of the countermeasures against those aforementioned physical layer attacks.

Eavesdropping Countermeasures: Eavesdropping cannot be prevented due to the wireless transmission nature. Signals can be easily intercepted with a receiver tuned to the transmission frequency. However, the confidentiality of the transmitted messages can be ensured. Encryption of the communication session could serve as a protection against the eavesdropping attack. Although the adversary is still able to intercept the transmitted signals between the legitimate reader and tag, he cannot recover any meaningful messages since he is unable to decrypt those signals.

Relay Attack Countermeasures: An important metric used to defend against the relay attack is the distance between the reader and the tag. The shorter the distance, the harder it is for the adversary to launch the attack. One of the most promising solution against relay attack is the distance bounding protocol proposed in [4, 17, 25, 20]. Here we will show one well known Hancke and Kuhn's distance bounding protocol [17] which could be used to detect the relay attack.

Their protocol is based on a challenge scheme. This is illustrated in Figure 5. The reader (verifier) and the tag (prover) share a secret K. The protocol is divided into two phases. *Phase 1* is a non-time critical phase. In this phase:

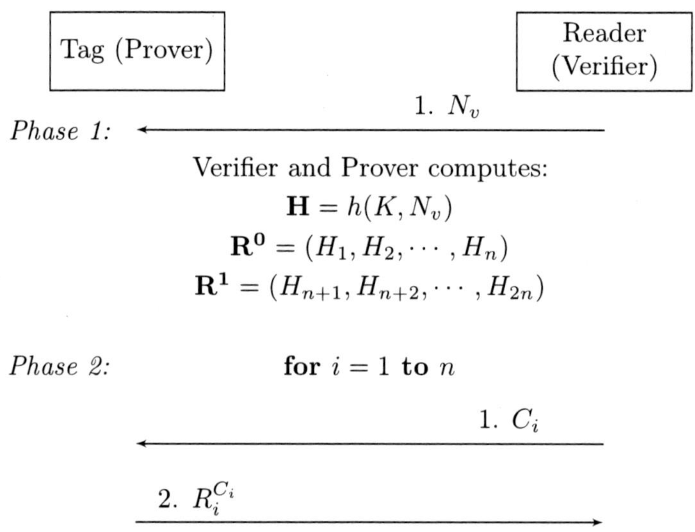

Figure 5. Hancke and Kuhn's distance bounding protocol.

1. The verifier sends a random number N_v to the prover, this random number is different for each session.

2. Both the prover and the verifier use the cryptographic function h, the secret K and the random number N_v to calculate two n-bit binary sequences $\mathbf{R^0}$ and $\mathbf{R^1}$ as follows:

$$R_1^0 R_2^0 \cdots R_n^0 || R_1^1 R_2^1 \cdots R_n^1 := h(K, N_v)$$

Phase 2 is a time critical phase. Phase 2 begins after a predefined number of clock cycles after the verifier sends out the random number to the prover. In this phase:

1. The verifier generates a random binary sequence $\mathbf{C}$ of length n, he sends out 1 bit C_i at a time for $1 \leq i \leq n$ to the prover.

2. The prover instantly replies with $R_i^{C_i}$.

3. The prover checks with his computed $R_i^{C_i}$. If the correct response is received within the sufficiently short time by the verifier, then the verifier accepts the prover as the valid tag. If the received response has exceeded the allowed time, a possible relay attack is alerted and the tag is deemed invalid.

The distance bounding protocol used to prevent the relay attack is highly sensitive to the timing of the signals. If the adversary tries to relay the signal to a tag outside of the reading range, the time takes would likely to be much longer. Once the response time by the tag exceeds the allowed time, the authentication would become unsuccessful. This is especially the case when there exists a fake prover and a fake verifier which is distant from each other, while trying to relay in between the legitimate reader and the tag.

This protocol is highly efficient and can protect against mafia fraud. However, it is still vulnerable against the terrorist fraud. Reid *et. al.* have proposed another efficient distance bounding protocol which can resist the terrorist fraud [25].

However, the distance bounding method is not without its weaknesses. First, all distance bounding protocols assume the adversary is further away from the reader than the tag is, this may not necessary be the case. Second, it has very high requirements on the clock synchronization and timing information for each round in phase 2. It further assumes the tag immediately returns the response of the reader's challenge bit. In reality, it takes several clock cycles before the tag can return the response. Moreover, the clock frequency of a typical RFID tag ranges from 1.92 MHz - 4 MHz [19, 18, 30]. Assuming the signals travels at the speed of light which is $3 \times 10^8 m/s$. This implies that one clock cycle delay in processing the data in the tag can result in an approximately 75 m - 200 m distance offset. Thus, determining the allowed time is a very difficult decision. Furthermore, the attacker with a much higher clock frequency could theoretically be used for relaying information and he may succeed in satisfying the timing requirement in a distance bounding protocol. Thus, the reliability of these systems are not questioned. Finally, this scheme requires a secure cryptographic function in order to generate two n-bit binary sequences $\mathbf{R^0}$ and $\mathbf{R^1}$.

Therefore, it cannot be easily implemented on an inexpensive passive RFID tag.

Other countermeasures against relay attack include the physical interaction or authentication of the tag, this include the push button configuration mechanism implemented in WiFi [29], the biometric authentication or the addition of password protections. However, all these methods would defeat the purpose of RFID, making it less convenient. To this date, there still not seems to be a perfect solution to the relay attack. This remains to be an interesting research problem.

Active Jamming Attack Countermeasures: The active jamming attack is applicable to most wireless communication systems not just for RFID systems. Conventional spread spectrum techniques are adopted to combat against jamming [24]. There are two types of spread spectrum communications: Direct sequence spread spectrum (DSSS) and frequency hopping spread spectrum (FHSS). In both cases, the reader and the receiver pre-share a sequence, it is called the spreading sequence in the DSSS case and the frequency hopping sequence in the FHSS case. In the DSSS scenario, each transmitted bit is expand in frequency by the employing spreading sequence, the received signals are collectively summed together to increase the radio signal strength by a factor of n, where n is the length of the spreading sequence. This n is also known as the processing gain. In the FHSS scenario, information are transmitted over different frequencies according to the pre-shared frequency hopping sequence in hopes that the adversary would not know which frequency to jam. Note that the FHSS has been incorporated in the EPCglobal Class 1 Gen 2 standard.

Disabling Tags Countermeasures: The physical tag destruction can be avoided with the better protection of the tag. Avoid placing the tag in the high electromagnetic field. In an environment with high radio interference and electromagnetic field, a shield can be used. Moreover, use an effective password protection for accessing the "KILL" command. Such as the 32-bit "KILL" password implemented in the EPCglobal Class 1 Gen 2 standard. Furthermore, making sure the password does not get disclosed to the unwanted parties. Without the password, the adversary cannot access the "KILL" command to permanently disable on the tag.

3.3. Network/transport Layer Attacks

The attacks in this layer exploit vulnerabilities on the communications between the legitimate reader and the tag. It also include exploiting weaknesses in the design of the communication protocols. In the network/transport layer, the connection between the reader and database is always considered secure. This is because the connection between them is usually through secure and robust wired connections. Unauthorized access to the communication session is extremely difficult if not impossible.

Figure 6 lists different attacks in the network/transport layer. Attacks in this layer can be further classified into three categories. Attacks on the reader, attacks on the tag and network protocol attacks. The attack we show in the previous section on the authentication protocol of EPCglobal Class 1 Gen 2 standard is an example of network/transport layer attack. Tag attacks further include tag cloning and spoofing. The reader attack includes impersonation of the reader.

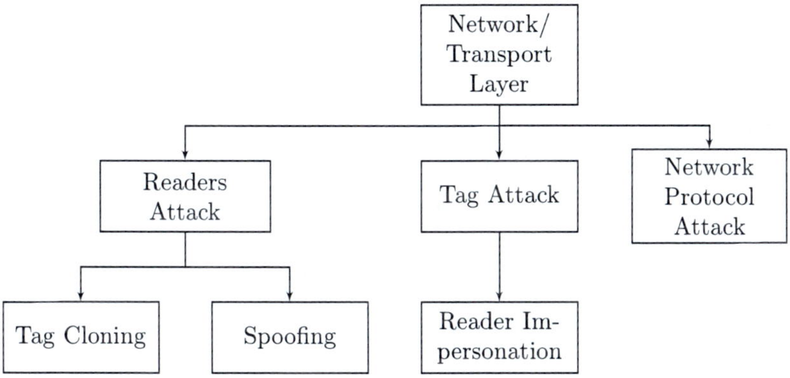

Figure 6. Network/Transport Layer Attacks.

Tag Cloning. This is also sometimes referred to as tag counterfeiting. Although each RFID tag should have an UID number, this is not always the case in reality as the attacker may try to clone or counterfeit a tag. In the case where RFID tags do not employ any security features as with most of the EPCglobal Class 1 Gen 2 tags, then cloning of the tag involves just copying the tag's ID along with its associated data to the cloned tag. The cloned tag would cause confusions as more than one tag with the same ID is circulated in the system. Thus, the integrity of the system is compromised. Even if the tag contains some security features, it is still susceptible this attack.

A practical example of tag counterfeiting is the the construction of an RF tape recorder by an undergraduate student reported in [28]. This device managed to read the commercial proximity cards primarily used for building access control, and simulate the signals to compromise the system and gain access to entering the building.

The feasibility of cloning the tag would depend on the level of security each tag possesses. In the case of active and semi-passive tags, reasonably secure features can be implemented to prevent tag cloning. However, in the case of constrained passive RFID tags, preventing tag cloning can be a challenging task.

Spoofing Attack: Spoofing attack is a variant of the tag cloning attack. However in this attack, the adversary does not physically clone an RFID tag. Instead, the adversary gains access to the legitimate RFID tag's information including the secret keys. He then tries to impersonate the legitimate tag when the reader makes the query by presenting the reader with the legitimate tag's credentials. This attack is very similar to the tag cloning attack in that both attacks require gaining privileges of the legitimate tag. Consequently, the countermeasures for both attacks are expected to be quite similar as well.

Reader Impersonation: This attack is analogous to spoofing attack on the reader side. In this attack, the unauthenticated adversary counterfeits the identity of the legitimate reader. He then performs the standard communication with the legitimate tags to fulfill his purposes. This attack is somewhat easier to prevent because of the capability of the reader. Compared to RFID tags, readers are much more powerful in terms of computing power and memory storage. More sophisticated security features can be implemented to thwart the reader impersonation attack.

Network Protocol Attack: The adversary in this scenario exploits the weaknesses in the design of the network protocol. This include but not limited to the reuse of the same random number, or in the case of the EPCglobal Class 1 Gen 2 standard, the back-to-back transmission of bitwise XOR of the random number and the key. All these design flaws can be exploited by the malicious adversary to launch attacks to compromise the corresponding RFID system.

3.4. Network-transport Layer Attack Countermeasures

To prevent the aforementioned reader and tag attacks, the mutual authentication between the reader and tag need to be established first. Two conventional methods that immediately come to mind are symmetric key and public key cryptosystems. These two methods are practical if both the reader and the tag are reasonably powerful. Arguably, this may be the case for the high end, more expensive active and semi-passive tags. These two types of tag are more powerful, conventional symmetric key or public key cryptosystems can be implemented to achieve mutual authentications. However, this is certainly not the case for the passive RFID tags. As we have discussed earlier, the inexpensive passive RFID tags are power, memory and computationally constrained, conventional public key and hash function options are not available to this type of tags. Conventional symmetric key authentication schemes are also not secure mainly due to insufficient key lengths. Passive RFID tags are the hardest ones to be kept secure from the malicious adversary. As a result, many alternative attempts have been made in the literature to achieve secure mutual authentications between the reader and the tag. We will examine a number of authentication schemes in this section.

Cloning Attack Countermeasures: Tuyls, Batina in [27] and Devadas *et al.* in [5] have proposed the use of a physical un-clonable function (PUF) to authenticate tags. A PUF is a function that maps a challenge to the corresponding response with the assist of physical hardware components. It must have the two following properties:

1. Easy to evaluate. This guarantees that the response can be generated from the challenge relatively easily.

2. Hard to characterize. This implies that the adversary who does not possess the device cannot predict the response to a challenge even if he can manage to gather a number of measurements.

PUFs can be viewed as a way to mimic the secure hash function. PUFs can be implemented in various ways. One of which is based on the timing and delay information of the integrated circuits (ICs) [9]. The manufacturing process causes significant delay differences among the different ICs even with identical circuit layouts. This is like the fingerprint of the IC. The PUF response can be derived from the delay and timing characteristics of the IC chip.

When the reader queries the tag, the tag is only authenticated if the PUF on the tag returns the correct response. One drawback of this approach is that there will be *intra PUF variations* and *inter-PUF variations* [5]. *Intra-PUF variations* refer to the reproducibility of the same PUF-circuit response. *Inter-PUF variations* measure the uniqueness of each individual PUF-circuit response. There are many factors that can affect the outcomes of the authentication results. This include the temperature and the environment. False positive caused by *intra-PUF variations* and false negative caused by *inter-PUF variations* will occur. This would affect the accuracy of the tag authentication. The reliability this scheme is questioned.

Similar to the idea of PUF, an RFID fingerprinting scheme is proposed in [23]. Rather than using the fingerprint on the timing and delay information of the IC, this scheme uses the fingerprint of the tag response. It is based on the claim that each RFID tag has its unique frequency responses. This unique frequency response can be served as the "fingerprint" for these tags. The reader checks the tag's frequency response during the authentication for the detection of cloned tags. The experimental results have shown there exist a considerable level of *false positives* and *false negatives*. Moreover, the frequency response of each tag varies considerably under different temperatures, humidity, *etc*. Although this scheme is very simple and requires practically no extra cost, the accuracy and repeatability is not guaranteed. The reliability is expected to be even worse than the PUF scheme, making this scheme highly impractical.

Database controlled cloning detection method has also been proposed [21]. The behaviour of each tag is temporarily stored in the back end server. This cloning detection is alerted when the tag behaves abnormally. For example, someone carrying a system access badge enters the building at 10:00AM. Ten minute later, another person trying to enter the building with the badge having exactly the same credentials. At the same time, the server observes that badge has never exited the building. Therefore, one potential case of badge cloning is detected. One drawback of this is that even though the system can detect the cloning, it cannot tell which one is the legitimate tag and which one is the cloned tag. Moreover, this detection scheme is not applicable to all scenarios.

One training based reader tag mutual authentication scheme has been proposed in [1]. The authors assume that only the legitimate reader and tag can always stay in the vicinity of each other. This assumption may be true in some applications such as in a home environment. Each legitimate tag is always in the reading range of the reader. The cloned tag or reader brought in by the adversary cannot always stay in the vicinity of the reader or tag. In this scheme, each tag will leak a small portion of its secret to the reader only after a certain amount of "quality" time have been spent by the reader and the tag. When the total required time has elapsed and link between the reader and the tag is still alive, then the reader authenticates the tag. At any point during this time, if the connection between them is lost, then the whole process has to start over again from the beginning. Therefore, by spending less than the threshold amount of time, the mutual authentication cannot be established.

Spoofing Attack Countermeasures: The spoofing attack could be prevented using most of authentication schemes listed in the countermeasure against tag cloning. Furthermore, an additional layer of security such as password or biometric authentications can be used to combat against this attack. However, by the same reasoning, the advantage of convenience of RFID is diminished.

Reader Impersonation Attack Countermeasures: Both the reader authentication and mutual authentications could prevent the reader impersonation attack. Note again that mutual authentications contain both the reader authentication and the tag authentication. The training based mutual authentication scheme introduced earlier can be used to combat against reader impersonation attack. Additionally, a second form of PIN or password protection can also thwart this attack. Again, this eliminates the advantage of the RFID system.

Network Protocol Attack Countermeasures: This attack can be prevented with the secure design of the network protocol between the reader and the tag, such as never reuse the random number, avoid simple exclusive or operation. Disabling insecure or unused network protocols running on the system. All of these would reduce the possibility of launching a network protocol attack by the adversary.

3.5. Application Layer Attack

The attacks on this layer target the vulnerabilities in the software applications. These attacks include unauthorized reading of tag contents, modification of tag data and middleware attacks. This is shown in Figure 7.

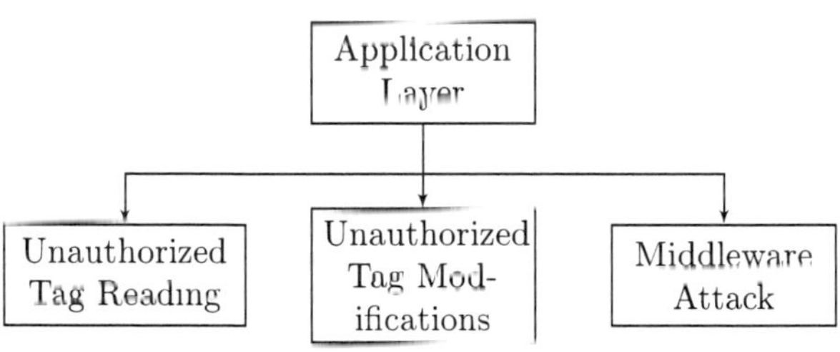

Figure 7. Application Layer Attacks.

Unauthorized Tag Data Reading and Modifications: Unlike most electronic products, RFID tags do not have an on/off switch. Any readers within the reading distance of the tag would be able to query the tag's information. If there are no security features implemented or security features are inadequate, then the adversary could easily gain access to the tag and read the contents stored in the tag. The privacy of the tag is compromised.

If the misbehaved reader can manage to read a tag's information, it is likely that he can also erase the original data contents inside the tag's memory and write his own data to the tag. Being able to modify the tag's content without being detected could potentially have serious implications on the RFID system. This is especially the case if the application of the RFID tag is being used for health and medical purposes. Patient's health and life could be put at the risk if the critical information in the tag is being altered.

Middleware Attacks: Middleware attack generally is a variant and more serious form of unauthorized tag modifications. In this case, the adversary's intention is no longer to just modify the contents stored in a tag, but rather by taking over the control of a single tag, the adversary can harm the entire RFID network. Two of most common middleware attacks are *buffer overflow attack* and *malicious code injection attack.*

- *Buffer Overflow Attack:* Buffer overflow attack is known to be a very hard security problem in any software applications. It exploits the data or code beyond the bounds of the fixed-length buffer. The adversary simply takes over the control of a RFID tag, repeatedly send some code to the reader untill it overflows the buffer in the back end RFID middleware.

- *Malicious Code Injection:* RFID can be used as a transport medium for infecting other entities including the reader and the back end database. Rieback *et al.* have demonstrated this attack in [26]. This attack can be devastating as injecting a small piece of code to one tag could potentially paralyze the entire RFID network.

3.6. Application Layer Attack Countermeasures

In this subsection, we will describe some of the common countermeasures against the application layer attacks.

Unauthorized Tag Data Reading and Modification Countermeasures: To defend against unauthorized tag data reading and modifications, access control mechanism to RFID tags should be implemented. The easiest form of solution is a read-only tag to combat the unauthorized content modifications. Another commonly adopted approach is the use of an aluminum-lined wallets to protect RFID tags from being misread by the unauthorized parties. However, this approach does not solve the problem because the tag can still be read once it is taken out of the aluminum-lined wallet.

Encryption techniques and properly designed authentication protocols could also serve as protections against unauthorized tag data reading and modifications. This include conventional symmetric key and public key schemes for encryption and hash functions for authentication. However, in the case of power and computational constrained passive tags, these approaches are not feasible. Alternative forms of protection are needed. We have already discussed possible alternatives in Section 3.4, network-transport layer attack countermeasures.

Middleware Attack Countermeasures: Buffer overflow and malicious code injection can be prevented with performing regular and extensive code review and rigorous checks to ensure the application is secure against any vulnerabilities and bugs. For database securities, the use of bound parameters and applying least possible privileges suggested by Friedl [8] could be applied to secure the database. Other measures such as turning off unnecessary protocols in the network, checking the input data.

3.7. Multi-Layer Attacks

Many attacks on the RFID systems cannot be confined to a specific layer. The adversary exploits the weaknesses in multiple layers to launch an attack. This include the covert channel attack, the traffic analysis, side channel attacks and the replay attack as shown in Figure 8. The multi-layer attack is generally considered as the strongest form of the attack.

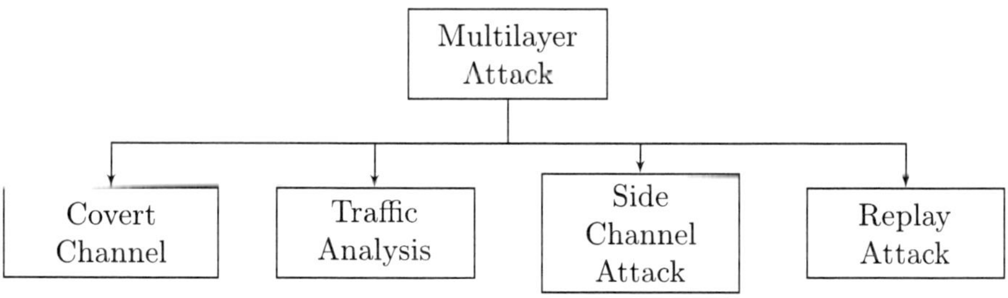

Figure 8. Multi-Layer Attacks.

Covert Channels: Adversary tries to create an unauthorized communication channel to gain access or to transfer data to the tag in a manner which is difficult to detect [12]. An example would be an adversary gains unauthorized access to the tag which has been implanted in a patient's body. Consequently, he can illegally retrieve's information regarding the patient's medical data.

Traffic Analysis: This attack is similar to the eavesdropping and it is applicable to any wireless communication systems. An adversary can intercept the communication between the reader and the tag and tries to extract useful information from the intercepted communications. Even if the communication between the reader and the tag is encrypted, it is still vulnerable to this attack.

Side Channel Attacks: Side channel attacks take advantages of physical implementation flaws of the cryptographic algorithms. The four main forms of side channel attacks are the timing attack, the fault attack, power analysis attacks and electromagnetic (EM) attacks. Power analysis / EM attacks can be further classified into the simple power analysis (SPA) / simple EM attack and the differential power analysis (DPA) / differential EM attack respectively.

- *Timing Attack:* The time attacking is first introduced by Kocher in [13]. It exploits the timing differences in the cryptographic computations. Usually implementations of cryptographic functions are performance optimized. The run time of a cryptographic function would to some degree depends on the key. A simple example is the conditioning branches during cryptographic operations. Based on the value of the key, the cryptographic operation either enters or not enters the conditioning branch. This causes a slight timing difference during the cryptographic computation. By measuring the time elapsed for the cryptographic operation, the adversary could guess which branch is taken. Thus, he can recover the value of the key.

- *Fault Attack:* Most cryptographic devices assume they can operate reliably and behaves exactly as expected under all circumstances when they are used. However, this is a strong assumption which can never prove to be true. Fault attacks on cryptographic primitives have been extensively studied since 1996 [3]. Hardware faults and errors can occur during the operation of a cryptographic function. This fault could be exploited by the adversary in recovering the secrets. The feasibility of fault attack would depend on the capabilities of the adversary and the type of faults it can generate. The fault attack is hardware implementation dependent. It is unique for different implementations of the cryptographic primitive.

- *Power Analysis Attack:* Power analysis attack can be further divided in simple power analysis (SPA) and differential power analysis (DPA) attacks [14]. The SPA gains insight on keys based on the power consumption variations of cryptographic operations. In the SPA attack, the adversary needs to have the knowledge of the implementation. This is different from the case of DPA. DPA is considered the most power side channel attack. DPA primarily uses statistical analysis and error-correction method to extract information on the keys.

- *EM Attack:* Similar to the power analysis attack, EM attacks can also be divided into the simple EM attack and the differential EM attack. They are analogous to SPA and DPA in the power analysis attack. The only difference is instead of exploiting the variations in power consumption, EM attacks exploit the information leaked from the electromagnetic field.

Replay Attacks: The adversary attempts to record one valid communication session's interchanged messages between the reader and the tag. He will then reuse these messages as his credentials in a later communication session to bypass the security function and establish a link with the tag or the reader.

3.8. Multi-Layer Attack Countermeasures

In this subsection, we will present some of countermeasures against multi-layer attacks described in the previous section.

Covert Channel Countermeasures: The covert channel attack is hard to defend against because by definition, the reader and the tag have no knowledge of unauthorized communications. A possible prevention to this attack is to reduce the availability of un-necessary memory resources.

Side Channel Attack Countermeasures: The key to preventing side channel attacks generally include [16].

- De-correlate output traces for each cryptographic runs.

- Careful selection of assembler instructions such that the timing information, power consumption and the EM radiation is hard to analyze.

- Make algorithmic changes to the cryptographic primitives to avoid creating timing differences and power consumption variations. Avoid using conditional branches in the cryptographic computations. Be careful when conducting performance optimizations.

To achieve the above goals, the following methods are proposed.

- *Randomization:* This is the most common method to counter side channel attacks. Data are randomized to avoid the leakage of timing information, power consumption and the EM radiation. The adversary cannot gain any useful information even if he performs different side channel attacks. The security level of the system is dependent on the randomization function.

- *Masking:* The data masking technique is a very popular technique used against power analysis and timing attack. The idea of masking is simple. The message and the key are hidden by a masking function before the computation starts. Adjustments are made at the end to recover the correct data. Similar to the randomization technique, the security is also dependent on the masking function.

- *Shielding:* The device is put in a shielded box to prevent adversaries from being able to measure the EM field and power consumption of the device. However, this solution is not effective if the adversary is able to remove the shield.

Traffic Analysis Attack Countermeasures: Traffic analysis cannot be fully prevented because of the nature of wireless communication. However, the use of encryption on the transmitted data could mask the message contents from the adversary. Without the decryption key, the intercepted signals become meaningless. This can alleviate the attack.

Replay Attack Countermeasures: The replay attack can be prevented with the use of timestamps, one-time passwords. Moreover, authentication protocols should be designed such that the random number is never reused. This ensures the response to the challenge would be different for every protocol run. Intercepting one round of communication by the adversary provides no benefit in the later rounds.

Conclusion

RFID technology has become very popular in the recent years. The tiny, inexpensive RFID tags can be attached to objects and even person for seamless identification purposes. It has simplified millions of people's life and revolutionized the way the world operates. Despite all these great advantages the RFID technology has offered us, it has also left us with numerous concerns which cannot be overlooked. One of the most important of which is the security and privacy concern. If not being dealt with properly, privacy of millions of people could be compromised. The need to keep RFID tags at a very low cost and at the same time, can withstand all known attacks is an extremely challenging task. Especially in the case the of passive RFID tags, conventional strong cryptosystem cannot be implemented, alternative methods are sought to defend against the adversaries.

In this chapter, we have given a brief overview on the security aspects of the RFID system. More specifically, we have shown some of the common attacks on the RFID system. These attacks are categorized by the different layers. Some existing countermeasures in the literature against these attacks are also presented. These countermeasures by no mean perfect solutions those attacks. Many research effort are put into securing different RFID systems. RFID security remains to be a hot research topic for the years to come.

References

[1] G.T. Amariucai, C. Bergman, and Y. Guan, An automatic, time-based, secure pairing protocol for passive RFID. In *Proceedings of RFIDSec*, pp. 108–126, 2011.

[2] G.Avoine and M.A. Bingo, S. Kardas, C. Lauradoux and B. Martin, A frqmework for analyzing RFID distance bounding protocols. *Journal of computer security*, vol. 19, no. 2, pp. 289–317, Apr. 2011.

[3] D. Boneh, R.A. DeMillo and R.J. Lipton, On the importance of checking cryptographic protocols for faults. In *Proceedings of CRYPTO '97*, LNCS 1233, pp. 37–51. 1997.

[4] S. Brands and D. Chaum, Distance-bounding protocols. In *Advances in Cryptology EUROCRYPT '03*, LNCS 765, pp. 344–359. Springer-Verlag, 1993.

[5] S. Devadas, E. Suh, S. Paral, R. Sowell, T. Ziola, and V. Khandelwal, Design and implmentation of PUF based "unclonable" RFID ICs for anti-counterfeiting and security applications. In *Proceedings of the 2008 IEEE international conference on RFID*, pp. 58–64, San Jose, USA, Apr. 2008.

[6] EPCGlobal, UHF class 1 gen 2 standard v. 1.2.0. 2008.

[7] L. Francis, G. Hancke, K. Mayes and K. Markantonakis, Practical NFC peer-to-peer relay attack using mobile phones. In *Radio Frequency Identification: Security and Privacy Issues*, pp. 35–49. Springer Berlin Heidelberg, 2010.

[8] S. Friedl, *SQL injection attacks by example* [Online], Available: http://www.unixwiz.net/techtips/sql-injection.html.

[9] B. Gassend, D. Clarke, M. van Dijk and S. Devadas, Silicon physical random functions. In *Proceedings of CCS '02*, pp. 148–160. 2002.

[10] A. Grover and H. Berghel, A survey of RFID deployment and security issues. *Journal of information processing systems*, vol. 7, no. 4, pp. 561–580, Dec. 2011.

[11] A. Juels, RFID security and privacy: A research survey. *Journal of Selected Area in Communication (J-SAC)*, 24(2):381–395, 2006.

[12] A. Karygicmnis, T. Phillips, and A. Tsibertzopoulos. RFID security: A taxonomy of risk. In *Communications and Networking in China, 2006. ChinaCom '06. First International Conference on*, Oct. 2006.

[13] P. Kocher, Timing attacks on implementations of Diffie-Hellmann, RSA, DSS,and other systems. In *Proceedings CRYPTO '96*, LNCS 1109, pp. 104–113. 1996.

[14] P. Kocher, J. Jaffe and B. Jun, Differential power analysis. In *Proceedings of CRYPTO '99*, LNCS 1666, pp. 388–397. 1999.

[15] G.P. Hancke, and M.G. Kuhn, An RFID distance boudning protocol. In *Proceedings of IEEE security and privacy for emerging areas in communication networks*, pp. 67–73. California, 2005.

[16] J. Goubin and J. Patarin, DES and differential power analysis. In *Proceedings of CHES '99*, LNCS 1717, pp. 158–172. 1999.

[17] G. Hancke and M. Kuhn. An RFID distance bounding protocol. In *Proceedings of the 1st international conference on security and privacy for emerging area in communication networks*, pages 67–73, Las Vegas, USA, 2005.

[18] Y. Hong, C.F. Chan, J. Guo, J.S. Ng, W. Shi, L.K. Leung, K.N. Leung, C.S. Choy and K.P. Pun, Design of passive UHF RFID tag in 130nm CMOS technology. In *Proceedings of IEEE Asia Pacific conference on circuits and system*, pp. 1371–1374. Maco, Dec, 2008

[19] Impinj. Gen 2 tag clock rate - what you need to know, 2006.

[20] C.H. Kim and G. Avoine, "RFID distance bounding protocols with mixed challenges," *IEEE Transaction on Wireless Communication*, vol. 10, no. 5, pp. 1618–1626, May. 2011.

[21] L. Miroski and J. Harnett, Deckard: A system to detect change of RFID tag ownership. *International Journal of Computer Science and Network Security*, vol. 7, no. 7, no. 89–98, 2007.

[22] A. Mitrokotsa, M.R. Rieback, and A.S. Tanenbaum, Classifying RFID attacks and defenses. *Information Systems Frontiers*, vol. 12, no. 5, pp. 491 - 505, Nov 2010.

[23] S.C.G. Periaswamy, D.R. Thompson, and J. Di, Fingerprinting RFID tags. *IEEE Transactions on Dependable and Secure Computing*, vol. 8, no. 6, pp. 938–943, Nov. 2011.

[24] M.B. Pursley, *Introduction to Digital Communications*. Pearson Prentice Hall, 2005.

[25] J. Reid, J.M. Nieto, T. Tang and B. Senadji, Detecting relay attacks with timing-based protocols. In *Proceedings of the 2nd ACM symposium on*

Information, computer and communications security ASIACCS '07, pp. 204–213. Singapore, 2007.

[26] N, Ruebacjm B, Crispo and A. Tanenbaum, Is your cat infected with a computer virus? In *Proceedings of IEEE International Conference on pervasive computing and communications*, pp. 169–179. Pisa, Oct, 2008.

[27] P. Tuyls and L. Batina, RFID tags for anti-counterfeiting. In *Proceedings of the cryptogrpher's track at the RSA conference 2006*, 3860, pp. 115–131, San Jose, USA, 2006. Springer.

[28] J. Westhues, Hacking the prox card. In S. Garfinkel and B. Rosenberg, editors, *RFID: Applications, Security and Privacy*, pp. 291–300. Addision-Wesley, 2005.

[29] WiFi Allicance. *WiFi protected setup specification*, version 1.0h, 2006.

[30] L.Q. Zhang and N. Wu, Power and energy optimizsation of the digital core of a Gen2 long range full passive RFID sensor tag. In *Proceedings of IEEE International Conference on RFID*, pp. 125–133. Beijing, Oct, 2008.

In: Radio-Frequency Identification (RFID) ISBN: 978-1-63117-750-7
Editor: Xinxin Fan, pp. 119-152

Chapter 6

LIGHTWEIGHT CRYPTOGRAPHY FOR PASSIVE RFID SYSTEMS: A RESEARCH SURVEY

***Kalikinkar Mandal**and Xinxin Fan*†**
Department of Electrical and Computer Engineering
University of Waterloo, Waterloo, ON, Canada

Abstract

With the rapid development of Radio Frequency Identification (RFID) technology, security and privacy in RFID systems have become critical issues for many applications. Due to the tight cost and extremely constrained resources of passive RFID tags, traditional cryptographic primitives designed for full-fledged computers might not be suited for securing RFID systems. There is a need of lightweight cryptographic primitives that can perform strong authentication and encryption, and provide other security functionalities for low-cost RFID tags. During the past few years, a multitude of lightweight cryptographic primitives such as lightweight block ciphers, stream ciphers, and pseudorandom number generators (PRNGs) have been designed for providing confidentiality and authenticity for passive RFID tags. In this chapter, we provide a research survey on recently proposed lightweight cryptographic primitives. For each cipher design we provide an overview of the description about the

*E-mail address: kmandal@uwaterloo.ca
†E-mail address: x5fan@uwaterloo.ca

design, implementation, and existing attacks. Moreover, we also conduct an extensive comparison among different primitives in term of their design and performance.

Keywords: RFID, security, privacy, lightweight cryptographic primitive

1. Introduction

The advent of ubiquitous computing has brought us into an era where Radio Frequency Identification (RFID) systems are becoming pervasive and is slowly replacing its old ancestor, i.e., the barcode. An RFID system typically consists of three components: RFID readers, RFID tags and a back-end server. Each RFID tag, also known as a transponder, contains a unique identifier and transmits data over the air using radio frequency in response to interrogation by an RFID reader. A back-end server can retrieve the detailed tag information based on the tag's response from a database which the server manages. While RFID technology stimulates numerous great applications such as access control, contactless payment, supply chain management, real-time item localization, etc., it also raises a lot of economical, security, and privacy related concerns [41], especially in sensitive application areas such as finance, healthcare, military, etc. Traditional cryptographic primitives that are designed for general computers and network servers are quite difficult to fit into low-cost RFID tags, due to the associated constraints with respect to computing capability, execution time, area, and power consumption. Those challenges have led to a new research direction of cryptography, commonly called *Lightweight Cryptography* [61], which is aimed to develop more efficient cryptographic architectures and implementations for resource-constrained smart devices like low-cost RFID tags.

There are a number of challenges that need to be kept in mind when designing lightweight cryptographic primitives for passive RFID systems:

- The cipher should have a simple architecture since low-cost RFID tags have limited logic functionality (i.e., one or a few finite state machines) and limited memory (i.e., a few Kbits);
- The cipher should occupy small area on an RFID integrated circuit due to the tight cost and large volume of RFID tags for mass deployment;

- The cipher should consume little power on RFID tags since they are passive devices and the power harvested from an RFID reader is quite limited;
- The cipher should have an appropriate key size and the resulting ciphertext size because of the limited communication bandwidth in passive RFID systems;
- The cipher should be performed reasonably fast (i.e., in the order of microseconds or milliseconds) due to the regulatory limitations.

All the aforementioned challenges need a designer to make appropriate tradeoffs among security, performance, and cost when conducing cipher design for passive RFID tags. For evaluating the suitability of a lightweight cipher design for RFID tags, the standard unit of measure called Gate Equivalent (GE)[1] has been widely used. For low-cost RFID tags, it is commonly agreed that no more than 2000 GE can be dedicated to security [40]. To date, the most compact serialized, encryption-only hardware implementation of the Advanced Encryption Standard (AES) needs 2400 GE[56]. The shortcomings of classical cryptographic primitives to meet the stringent requirements of passive RFID systems have resulted in the rise of quite a few new proposals with innovative ideas during the past few years. Those proposals include lightweight symmetric-key cryptographic primitives (i.e., block ciphers, stream ciphers, hash functions, and pseudo-random number generators) and lightweight implementations of classical public-key cryptographic primitives (i.e., elliptic curve cryptography, lattice-based NTRU, etc.). While lightweight public-key primitives might have advantages for key management in RFID systems, the required resource for public-key primitives is much larger than that of symmetric-key counterparts.

In this chapter, we present a comprehensive research survey for recently proposed lightweight symmetric key primitives, with a focus on block ciphers in Section 2, stream ciphers in Section 3, and pseudorandom number generators in Section 4. The design principle of each lightweight symmetric-key primitive is detailed, followed by the description of the existing cryptanalysis results as well as the implementation results on various hardware platforms (e.g., FPGAs or ASICs). A comparison of hardware implementations of lightweight symmetric-key primitives is also presented at the end of each section.

[1]One GE refers to the area taken by one NAND Gate with two inputs and one output in Complementary metal-oxide-semiconductor (CMOS) technology.

2. Lightweight Block Ciphers

This section presents several lightweight block ciphers developed in recent years for RFID tags and wireless sensor nodes. Most of the designs are aimed to reduce the hardware complexity and power consumption and to increase the throughput. For each block cipher, we provide a design overview and the existing significant attacks, and discuss the hardware implementation cost. All the presented block ciphers take a plaintext of 64 bits as input and output a ciphertext of 64 bits except the block cipher CLEFIA.

2.1. Block Cipher: DESL and DESXL

DESL, proposed by Leander et al. [44], is a lightweight variant of the Data Encryption Standard (DES), which takes a 64-bit plaintext as input and outputs a 64-bit ciphertext where the key length is of 64 bits. The DESL is obtained from the original DES by replacing eight different S-boxes by a single cryptographically strong S-box with improved nonlinearity and by omitting the initial permutation step and its inverse permutation step as they do not provide any cryptographic strength. The S-box is chosen so that it is resistant to the linear approximation attack, the differential attacks, and Davis-Murphy attack [44]. That is, the design of DESL is based on the Feistel network where the round function is composed of one S-box. The main reason for replacing eight S-boxes by one S-box is to have a serial implementation in order to reduce the number of logic gates. DESXL is another lightweight variant of the DES. The difference between DESXL and DESL is that DESXL is based on the FX construction [63] and has the key whitening part, whereas DESL does not have the key whitening part.

In [44], the authors implemented DESL in serialized ASIC implementation and their implementation requires 1848 GEs. Note that the DESL implementation requires 20% less GEs than the DES implementation. To encrypt a plaintext of 64-bit, the implementation requires 144 clock cycles and the throughput reaches 5.55 Kbps. When the encryption algorithm runs at 100 KHz, the average current consumption is 0.89μA.

2.2. Block Cipher: mCrypton

Motivated by the design of the block cipher Crypton [46], Lim and Korkishko proposed a 64-bit block cipher mCrypton with secret keys of lengths 64, 96 and

128 bits for low-cost RFID tags and sensor nodes [47]. We denote an mCrypton instance with K-bit secret key by mCrypton-K. The goal of designing mCrypton is to come up with a block cipher with extreme efficiency in resource usages and power consumption and that can be hardware integrated or software implemented in tiny processor for resource-constrained devices. The encryption algorithm iterates the round transformation 12 times where the round transformation is composed of a nibble-wise substitution, a column-wise bit permutation, a column-to-row transposition, and a sub-key addition. The substitution process uses four S-boxes, and among four S-boxes, two S-boxes are inverse of the other two. Note that the encryption and decryption algorithms are the same except the key scheduling process.

The authors showed that the full-round, i.e., 12-round mCrypton is secure against linear and differential cryptanalyses [47]. In [59], Park proposed an 8-round related-key rectangle attack on mCrypton-128 with data and time complexity 2^{46}. Mala et al. [49] proposed a 9-round related-key impossible differential attack against mCrypton with 96 and 128 bits key. The data and time complexities of the attack on mCrypton-128 are $2^{59.7}$ and $2^{66.7}$, respectively, and the data and time complexities of the attack against mCrypton-96 are $2^{59.9}$ and $2^{74.9}$, respectively. Most recently, Shakiba et al. [65] developed an attack called *non-isomorphic bicliquie attack* on the full round mCrypton-96 and mCrypton-128. The time complexities of the non-isomorphic bicliquie attack with 4-round and 5-round bicliquies on mCrypton-96 are $2^{94.09}$ and $2^{93.75}$, respectively. Similarly, the time complexity of the non-isomorphic bicliquie attack with 4-round bicliquies on mCrypton-128 is $2^{126.05}$.

All variants of mCrypton-K with $K = 64, 96$, and 128 have been implemented using 0.13μm CMOS technology in [47]. The gate requirements for mCrypton-K with $K = 64, 96$, and 128 are 3473, 3789, and 4108 GEs, respectively, where the implementation costs include the costs of both encryption and decryption processes and the key scheduler [47]. The implementation cost for the encryption-only process can be reduced by at least 25%.

2.3. Block Cipher: KATAN & KATANTAN

The KATAN and KATANTAN family is a lightweight block cipher family based on feedback shift registers [16]. The family contains six variants in total, three variants KATAN-n for the KATAN family and three variants KATANTAN-n for the KATANTAN family with $n = 32, 48$, and 64. A cipher instance KATAN-

n or KATANTAN-n of the KATAN and KATANTAN family accepts a key of length 80 bits and a message of length n bits and it outputs an n-bit ciphertext after running the cipher 254 rounds. KATAN-n is composed of two registers with nonlinear internal state update functions where n is the sum of the lengths of two registers. The architectures of the KATAN family and the KATANTAN family are the same, except the key scheduling algorithm. In the KATAN family, round keys of KATAN-n are generated by using an LFSR of 80 bits where the characteristic polynomial of the LFSR is chosen to be primitive for achieving the maximum period. Whereas the key scheduling algorithm of the KATANTAN family is designed using a linear feedback shift register (LFSR) of length 8, and two multiplexers of 16-bit and 4-bit.

In [13], Bogdanov and Rechberger have developed a variant of the meet-in-the-middle attack to the full-round KTANTAN family by exploiting some weaknesses in the bitwise key scheduling algorithm. The time complexity of the attack for the full-round KATANTAN-32 is $2^{75.17}$ encryptions, for the full-round KATANTAN-48 is $2^{75.04}$ encryptions, and for the full-round KATANTAN-64 is $2^{75.58}$ encryptions, where for KATANTAN-32 the attack requires 3 plaintext/ciphertext pairs and for KATANTAN-48 and KATANTAN-64, the attack requires 2 plaintext/ciphertext pairs. Afterwards, Wei et al. [69] proposed an improved meet-in-the-middle attack to the full-round KTANTAN family. The improved attack has time complexities $2^{72.9}$, $2^{73.8}$ and $2^{74.4}$ against KATANTAN-n with $n = 32, 48$ and 64, respectively. Zhu and Gong [72] presented a multidimensional meet-in-the-middle attack to the reduced-round KATAN family and the time complexity for KATAN-32 with 175-round, KATAN-48 with 130-round and KATAN-64 with 112-round is $2^{79.30}$, $2^{79.45}$ and $2^{79.45}$, respectively.

There is an implementation tradeoff between the area and the throughput of the KATAN and KATANTAN family. For achieving the throughput 75.3 Kbps, KATAN-64 and KATANTAN-64 require 1269 GEs and 1168 GEs on 0.13μm CMOS technology, respectively [16]. For details about the implementation and the hardware cost of other instances, the reader is referred to [16].

2.4. Block Cipher: PRESENT

PRESENT is an ultra-lightweight hardware-oriented block cipher designed by Bogdanov et al. for providing confidentiality in constrained environments such as RFID tags and sensor networks [12]. It is one of the block ciphers in the ISO/IEC standard 29192-2 [38]. The design of PRESENT is based on a substitution-permutation network (SPN). Each round consists of a round-key addition layer, an S-box layer, and a permutation layer where the S-box layer uses a 4-bit S-box sixteen times. PRESENT accepts a message block of 64 bits and a key of length either 80 or 128 bits. After accepting a message, the cipher is iterated for 31 rounds to produce a ciphertext.

The authors of PRESENT performed a security analysis against various attacks including linear and differential cryptanalysis, truncated differential cryptanalysis, algebraic attacks, key scheduling attacks, and structural attacks. Nakahara Jr. et al. [57] proposed an algebraic cryptanalysis on 5-round PRESENT and the attack can recover half of the key bits within 2.5 minutes. They also proposed a key recovery attack based on a 21-round linear hull on 25-round PRESENT with 128-bit key with the entire codebook, which requires $2^{98.68}$ encryptions, 2^{40} blocks of memory. The success rate of the attack is 0.6. In [22], Collard and Standaert launched a statistical saturation attack against PRESENT and the attack can break up to 15-round of PRESENT using $2^{35.6}$ plaintext-ciphertext pairs. Özen *el al.* [58] presented a related-key cryptanalysis on PRESENT with 128-bit key and the complexity of the attack is as follows: The data complexity is 2^{63}, the memory complexity is 2^{53}, and the time complexity is 2^{104} memory access. In [68], Wang presented a differential cryptanalysis against 16-round PRESENT by finding and applying a 14-round differential characteristic with probability 2^{-62}. The data complexity of the differential attack is 2^{64} chosen plaintexts, 2^{32} 6-bit counters, and the time complexity is 2^{64} memory accesses.

Block cipher PRESENT was designed to achieve low hardware complexity. PRESENT with 80-bit key was implemented in VHDL and synthesized using the Virtual Silicon (VST) standard cell library based on the UMC L180 0.18μ process and the implementation requires using 1570 GEs [12]. According to the implementation in [12], it requires 32 clock cycles for encrypting a 64-bit message and the simulated power consumption is 5μW. Note that the throughout of PRESENT is 200 Kbps, which is higher than that of DESL. On the other hand, the gate equivalent required for implementing PRESENT with 128-bit key is 1886 GEs.

2.5. Block Cipher: PRINCE

PRINCE [14] is a hardware-oriented block cipher based on the FX-construction, which accepts a message of 64 bits and a key of 128 bits. The motivations behind designing PRINCE are to compute a ciphertext in a single clock cycle for a plaintext and to have moderate hardware costs. In the key expansion phase, a 128-bit secret key is expanded to a 192-bit key by applying a linear transformation in order to perform the key whitening technique to increase key strength. The entire encryption process consists of a key whitening technique and a PRINCE_{core} with 12 rounds where each round is composed of an S-box layer, a linear layer, a round constant addition layer and a key addition layer. PRINCE uses the α-reflection property to reduce the overhead of implementing the decryption algorithm over the encryption. In the encryption process, after the first five rounds, the S-box layer, another linear layer obtained from the linear layer of the round function and the inverse S-box layer are applied and then another five rounds are applied. The reason for involving another linear layer is to ensure the α-reflection property of the cipher.

In [14], the authors presented a security analysis of PRINCE considering linear and differential attacks, biclique and meet-in-the-middle attacks, algebraic attacks, and the slide attack. The authors claimed that the time complexity of recovering a secret key is at least 2^{127-n} when an adversary is able to obtain 2^n plaintext/ciphertext pairs. In [39], Jean et al. first performed a third-party cryptanalysis of PRINCE where they considered related-key boomerang attacks, related-key attacks with the α-reflection property, a single pair of related-key, memory-data tradeoff and time-memory-data tradeoff attacks on the full round PRINCE and PRINCE_{core}, and the integral attacks on the reduced 4-, 5- and 6-round of PRINCE and PRINCE_{core}. The related-key attack on the full-round PRINCE exploits the α-reflection property and the attack has a data complexity of 2^{33}, a time complexity of 2^{64}, and a memory complexity of 2^{33}. Again, the related-key boomerang attack on PRINCE_{core} is based on the related-key differential characteristic and one requires 2^{39} chosen plaintexts and 2^{39} memory to mount the attack. The time complexity of this attack is 2^{39}. It is a worth observation that a key recovery attack on PRINCE_{core} can be launched by modifying the values of α. The data and time complexities of the single-key boomerang attack with a chosen α are 2^{41}. Similarly, the time complexity of the single-key attack on PRINCE is $2^{125.47}$. The integral attack against 6-round PRINCE has the time complexity 2^{64}, and the data and memory complexities 2^{16}. The integral attack against 6-round PRINCE_{core} has the time complexity 2^{30}, and the data and memory complexities 2^{16}.

The hardware implementation cost of PRINCE is provided in Table 2. For the details of the hardware implementation, see [14].

2.6. Block Cipher: PRINT$_{\text{CIPHER}}$

At CHES 2010, Knudsen et al. proposed the block cipher PRINTcipher based on an SP-network for printable circuit, the IC-printing technology [42]. PRINTcipher has two variants PRINTcipher-48 with 80-bit key and PRINTcipher-96 with 160-bit key. PRINTcipher-n takes an n-bit plaintext as input and produces an n-bit ciphertext after iterating 48 and 96 rounds for $n = 48$ and 96, respectively. The encryption algorithm of PRINTcipher is composed of a key bitwise XOR, a fixed linear diffusion layer, a round constant generated by an LFSR and added using XOR, a keyed permutation layer and an S-box layer. The S-box layer uses four 3-bit S-boxes for introducing nonlinearity in the cipher and the same subkey is used for all rounds.

The authors claimed that PRINTcipher is secure against differential and linear cryptanalyses, higher order differential attacks, algebraic attacks, related-key attacks, and statistical saturation attacks [42]. Abdelraheem et al. [2] first launched two differential attacks on 22-round PRINTcipher-48, and the attack requires 2^{48} plaintext/ciphertext pairs for 22-round and the computational complexity of the attack is about 2^{48}. Leander et al. [3] proposed the invariant subspace attack against PRINTcipher. The attack exploits the fact that 2^{52} keys for PRINTcipher-48 and 2^{102} keys for PRINTcipher-96 are weak. For the weak keys, the attack can be considered as a distinguishing attack.

PRINTcipher variants are implemented in VHDL using *UMCL18G212T3* for measuring their performance in hardware. A compact hardware implementations of PRINTcipher-48 and PRINTcipher-96 in VHDL require 402 and 726 GEs, respectively [42].

2.7. Block Cipher: KLEIN

KLEIN is a lightweight block cipher based on an SP-network, which was designed to be very fast in the software platform and compact in the hardware platform and was targeted for wireless sensor nodes as well as RFID tags [31]. KLEIN takes a 64-bit plaintext as input and outputs a ciphertext of 64-bit where the key sizes are 64, 80 and 96 bits. An instance of KLEIN with key size K is denoted by KLEIN-K. The round transformation of the encryption process of KLEIN consists of four subroutines, namely AddRoundKey, SubNibbles, RotateNibbles, and MixNibbles, and the number of rounds required to obtain a ciphertext for KLEIN-64/80/96 is 12/16/20, respectively. The round keys are obtained by using a key scheduling algorithm based on a Feistel-like structure.

The SubNibble step is constructed using the same 16 S-boxes of 4-bit, which is the nonlinear components of the cipher, and other components such as RotateNibbles and MixNibbles are linear.

The authors showed that KLEIN is resistant to linear attacks, differential attacks, algebraic attacks, side-channel attacks and the integral attack on the key scheduling [31]. Aumasson et al. [9] proposed a practical chosen-plaintext key-recovery attack on up to 8-round of KLEIN-64 based on a collection of iterative differential characteristics. In [71], Yu et al. presented an integral attack on 7-round KLEIN-64 and truncated differential attacks on 8-round KLEIN-64 and differential attacks on 8-round KLEIN-80. However, the computational complexity for 8-round KLEIN-80 is $2^{77.5}$. Ahmadian et al. [5] developed a(n) (asymmetric) biclique attack on the full-round KLEIN-64 by exploiting weaknesses in the diffusion layer and in the key scheduling algorithm of KLEIN-64. The time complexity and data complexity of the attack are $2^{62.84}$ and 2^{39}, respectively.

To measure the performance of KLEIN in hardware, the authors use the VST standard cell library on the UMC 0.18 μm L180 Process, and the areas required for implementing KLEIN-64, 80 and 96 are 1981 GEs, 2097 GEs, 2213 GEs, respectively [31].

2.8. Block Cipher: CLEFIA

CLEFIA is one of the block ciphers in the ISO/IEC Standard 29192-2 [38]. CLEFIA, designed by Sony Corporation [73], is an iterative 128-bit block cipher, which accepts keys of 128, 196, and 256 bits and the numbers of rounds for three different key sizes are 18, 22, and 26, respectively [66]. CLEFIA uses a 4-branch and a 8-branch generalized Feistel networks as component functions. The cipher has two different round functions, which are constructed using two 8-bit S-boxes. The first 8-bit S-box is built by combining four 4-bit S-boxes and the second 8-bit S-box is designed using the inverse function over $\mathbb{F}_{2^8}$. Note that the design of the round function of the Feistel network is based on an SP-network, which has a round-key addition layer, an S-box layer and a permutation layer. The encryption and decryption functions use the 4-branch generalized Feistel network and the key scheduling for 192-bit key uses the 8-branch generalized Feistel network. This cipher uses the key whitening at the beginning and the end of the data processing stage. For the further details on the design we refer the readers to [66].

A number of cryptanalytic attacks such as integral attacks, impossible differential cryptanalysis on reduced-round CLEFIA and differential power analysis on CLEFIA can be found in [45, 48, 10, 62]. A summary of the various attacks against CLEFIA can be found in [15].

CLEFIA was designed to achieve good performance in hardware as well as in software. A hardware implementation in a 0.09μm CMOS ASIC library shows that CLEFIA with a 128-bit key can be implemented using 5979 logic gates at a throughput of about 1.60 Gbps [66]. A software implementation on Athlon 64 (AMD64) 4000+ 2.4 GHz processor shows that one can achieve a throughout 12.9 cycle/byte in the single-block implementation.

2.9. Block Cipher: SIMON and SPECK

Very recently, Beaulieu et al. [11] designed the Simon and Speck block cipher families based on the Feistel structure. Each family contains ten variants for different lengths of plaintexts and key sizes. The different lengths and key sizes of the cipher instances are provided in Table 1. A cipher instance of Simon is denoted by SIMON-$2n/K$ where $2n$ is the length of a plaintext and K is the length of the key. Encryption and decryption algorithms of Simon use simple operations such as bitwise XOR, AND and left circular shift. We note that the round function of Simon is composed of bitwise XOR, AND and left circular shift, where the AND operation is used for introducing nonlinearity in the cipher. On the other hand, the Speck family is designed using bitwise XOR, left and right circular shifts, integer addition, addition modulo 2^n, and modular subtraction. The round keys of the Simon family are generated recursively using an LFSR with input over a finite field $\mathbb{F}_{2^n}$, where the input sequences are generated by three other LFSRs of 5-stages. The feedback function for the LFSR over $\mathbb{F}_{2^n}$ is defined using the right circular shift and XOR operations.

In [6], Alkhzaimi and Lauridsen presented the differential attacks for key recovery and impossible differential attacks on all variants of reduced-round SIMON. For instance, for SIMON-64/96 with 24-round, the complexity of the differential attack is: time complexity $2^{58.42}$, memory complexity 2^{32} and data complexity $2^{62.01}$. Again, for SIMON-64/96 with 16-round, the complexity of the impossible of the differential attack is: time complexity $2^{91.98}$, memory complexity $2^{60.20}$, and data complexity $2^{65.24}$ [6]. Independently, Abed et al. [4] preformed the linear cryptanalysis, differential cryptanalysis, and rectangle attacks on the reduced-round SIMON family. Their security analysis is aimed

Table 1. Simon and Speck cipher family parameters

Block size ($2n$)	Key size	# of rounds	Block size ($2n$)	Key size	# of rounds
32	64	32	96	96	52
48	72	36	96	144	54
48	96	36	128	128	68
64	96	42	128	192	69
64	128	44	128	256	72

for the higher reduced-round of SIMON. For instance, the complexity of the differential attack on SIMON-64/96 with 26-round is: time complexity $2^{63.9}$, memory complexity 2^{31}, and data complexity 2^{63}.

Hardware implementation costs of different variants of the Simon and Speck families are provided in Table 2.

2.10. Summary of the Block Ciphers

This chapter presents hardware- and software-oriented lightweight block ciphers for resource-constrained RFID tags and wireless sensor nodes. Among these primitives, DESL, KATAN & KATANTAN, PRESENT, PRINCE, and $\text{PRINT}_{\text{CIPHER}}$ are purely hardware-oriented, and mCrypton, KLEIN, SIMON and SPECK are both hardware- and software-oriented. Note that the design of the block ciphers DESL, SIMON and SPECK is based on the Feistel structure, whereas the design of mCrypton, PRESENT, and $\text{PRINT}_{\text{CIPHER}}$ is based on the SP-network. In addition, the design of DESXL and PRINCE is based on the FX construction. One can observe that CLEFIA uses a Feistel network and an SP network for designing the encryption, decryption and key scheduling algorithms, and KLEIN employs an SP-network in the encryption and decryption algorithms and a Feistel network in the key scheduling algorithm. A summary of the performance of nine block ciphers are presented in Table 2.

3. Lightweight Stream Ciphers

Recently, many lightweight stream ciphers have been developed, for instance a few can be found in the eStream project [27]. The designs of the stream ciphers

Table 2. A performance comparison of the block ciphers

Cipher	Block (bits)	Key (bits)	Area (GEs)	Throughput Kb/s	Logic Process
DESL [44]	64	56	1848	44.4	0.18 μm
mCrypton [47]	64	64	2420	492.3	0.13 μm
KATAN-32/48/64 [16]	32/48/64	80	802/927/1054	12.5/18.8/25.1	0.13 μm
KATANTAN-32/48/64 [16]	32/48/64	80	462/588/688	12.5/18.8/25.1	0.13 μm
PRESENT-80 [12]	64	80	1570	200	0.18 μm
PRESENT-80 [64]	64	80	1000	11.4	0.35 μm
PRINCE [14]	64	128	3491	533.3	0.13 μm
PRINTcipher-48 [42]	48	80	402/503	6.25/100	0.18 μm
PRINTcipher-96 [42]	96	160	726/967	3.13/100	0.18 μm
KLEIN-64/80/96 [31]	64	64/80/96	1981/2097/2213	–	0.18 μm
CLEFIA [66]	128	128/196/ 256	5979/8536/ 8482	1605.49/1201.85/ 1016.95	0.09 μm
SIMON-48 [11]	48	96	763	15.0	0.13 μm
SIMON-64 [11]	64	96/128	838/1000	17.8/16.7	0.13 μm
SIMON-96/128 [11]	96/128	96/128	984/1317	14.8/22.9	0.13 μm
SPECK-48 [11]	48	96	884	12.0	0.13 μm
SPECK-64 [11]	64	96/128	984/1127	14.5/13.8	0.13 μm
SPECK-96/128 [11]	96/128	96/128	1134/1396	13.8/12.1	0.13 μm

to be presented in this section are based on linear feedback shift registers (LFSRs) and/or nonlinear feedback shift registers (NLFSRs). The main challenges in designing a stream cipher are to ensure that 1) The keystreams have large periods; 2) The ciphers should be resistant to the cryptanalytic attacks; and 3) The cipher should have efficient implementations on low-cost software and hardware platforms. Due to the lack of the theory of NLFSRs, in the current design methodologies, the LFSRs are commonly used for assuring the long periodicity of an output keystream.

3.1. Notations

Some notations are defined here for explaining the design of the stream ciphers.

- $\mathbb{F}_2 = \{0, 1\}$. the Galois field with two elements
- $\mathbb{F}_{2^m}$: a finite field with 2^m elements, which is defined by a primitive polynomial of degree m over $\mathbb{F}_2$

3.2. Stream Cipher: Grain Stream Cipher Family

Grain is a stream cipher family designed for the resource-constrained devices such as RFID tags. The design of the Grain family is based on a combination of a linear feedback shift register and a nonlinear feedback shift register. More specifically, an instance from the Grain stream cipher family consists of three fundamental blocks, namely an LFSR of length n, an NLFSR of length n and a filtering function. There are total three variants of the Grain family namely, Grain v0, Grain v1, and Grain-128 [33, 34, 27] and Grain v1 is one of the eStream profile 2 finalists [27]. In this section, we describe Grain-128 stream cipher of the Grain family. A general overview of the architecture of Grain-128 is provided in Fig. 1.

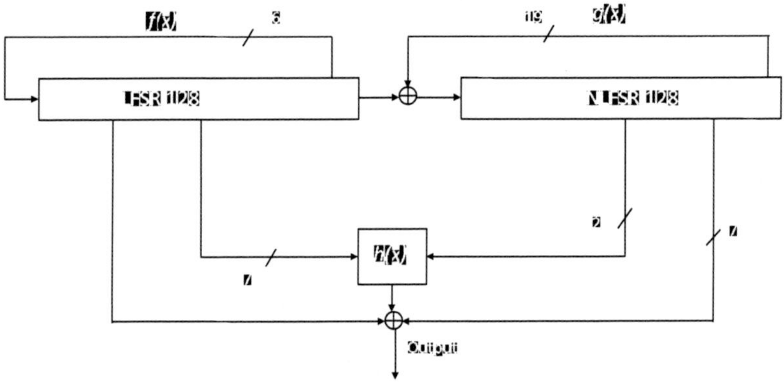

Figure 1. A block diagram of stream cipher Grain-128.

3.2.1. Design Specification of Grain-128

Grain-128 [34] is comprised of an LFSR and an NLFSR of length 128 bits and a filtering function in 17 variables. The key length and the IV length of Grain-128 are of 128 bits and 96 bits, respectively. The key bits are loaded into the NLFSR and the initial vector (IV) bits are loaded into the first 96 cells of the LFSR in the initialization phase and then the output bits are produced after 256 rounds of the initialization phase.

We denote by $(a_i, a_{i+1}, \cdots, a_{i+127})$ an internal state of the LFSR and $(b_i, b_{i+1}, \cdots, b_{i+127})$ an internal state of the NLFSR. The characteristic polynomial of the LFSR $f(x) = 1 + x^{32} + x^{47} + x^{58} + x^{90} + x^{121} + x^{128}$ is chosen to be primitive so that the LFSR can generate sequences of period $2^{128}-1$. The internal state of the LFSR is updated as follows

$$a_{i+128} = a_i + a_{i+7} + a_{i+38} + a_{i+70} + a_{i+81} + a_{i+96}, \ a_i \in \mathbb{F}_2, i \geq 0.$$

The internal state of the NLFSR is updated as

$$\begin{aligned} b_{i+128} = a_i + b_i + b_{i+26} + b_{i+56} + b_{i+91} + b_{i+96} + b_{i+3}b_{i+67} + b_{i+11}b_{i+13} + \\ b_{i+17}b_{i+18} + b_{i+27}b_{i+59} + b_{i+40}b_{i+48} + b_{i+61}b_{i+65} + b_{i+68}b_{i+84}, \ a_i, b_i \in \mathbb{F}_2 \end{aligned}$$

where the nonlinear feedback function is a sum of a linear function and a quadratic bent function. Note that the nonlinear feedback functions for another two variants (i.e., Grain v0 and Grain v1) of the Grain family are of degree 6. The output sequence of Grain-128, denoted by $\{z_i\}$, is obtained as follows

$$z_i = \sum_{j \in I} b_{i+j} + s_{i+93} + h(b_{i+12}, s_{i+8}, s_{i+13}, s_{i+20}, b_{i+95}, s_{i+42}, s_{i+60}, s_{i+79}, s_{i+95}),$$

where $h(x) = x_0x_1 + x_2x_3 + x_4x_5 + x_6x_7 + x_0x_4x_8$ is a function of nonlinearity 240, degree 3 and algebraic immunity 3, and $I = \{2, 15, 36, 45, 64, 73, 89\}$. We can observe that the output sequence is obtained by combining 9 internal state bits from the NLFSR and 8 internal state bits from the LFSR. The initialization phase of the Grain family is the same as the running phase except the output sequence is XORed with the feedback function of the LFSR and the NLFSR and no keystream is output during the initialization phase. The period of the keystream $\{z_i\}$ is bounded below by $2^{128}-1$.

As the long period of a keystream is a desirable property, the Grain stream cipher family can guarantee the lower bound of a keystream. Hu and Gong [36] proved that the period of a keystream of Grain is a multiple of $(2^n - 1)$, which is promised due to employing the LFSR component in the cipher. The design of the Grain family without the filtering function can be regarded as a composition of feedback shift registers or a nonlinear product of feedback shift registers [32]. In 1970, Green and Dimond studied the period property of the composition of an LFSR and an NLFSR and vice-versa for small parameters [32].

3.2.2. Existing Attacks on the Grain Stream Ciphers and Implementation

Various attacks on different variants of Grain stream ciphers can be found in [26, 7, 24, 55, 43, 30]. Dinur and Shamir [24] applied a key-recovery *dynamic cube attack* on Grain-128. According to the dynamic cube attack on the full-round Grain-128, the time complexity of the attack to recover the key is 2^{113} when it is assumed that the set of weak keys contains only 2^{10} of the 2^{128} possible keys. Dynamic cube attack on two reduced-round variants of Grain-128 is presented in [24]. Knellwolf et al. [43] proposed a conditional differential cryptanalytic attack based on differential cryptanalysis, which leads to a distinguishing attack and a key-recovery attack that can recover partial key bits. The attack is applied on reduced-round variants of Grain-128 and Grain v1. According to the conditional differential cryptanalysis, for a 104-round variant of Grain v1, it is possible to build a distinguisher with complexity 2^{39} when the distinguishing probability is 0.58, and for a 213-round variant of Grain-128, the attack can recover three key bits with probability up to 0.59.

Mihaljević et al. [55] presented an attack based on the normality of the filtering function for recovering the internal state of Grain v1. Since the normality of the filtering function of Grain v1 is 2, the expected time complexity and the memory complexity for recovering the internal state are about 2^{54} and 2^{70}, respectively, and an attacker requires 2^{34} keystreams for different IVs and each keystream is of length 2^{38}. Many distinguishing attacks can be found in [26, 7, 67], and a sliding property in the initialization phase of the Grain family is used to reduce the exhaustive search complexity by a half [19].

3.3. Stream Cipher: Trivium

Trivium, designed by De Canniére and Preneel, is a bit-oriented synchronous stream cipher designed for achieving high throughput and for providing security for resource-constrained environments [17, 18]. It is also one of the eStream profile 2 finalists [27]. The length of the internal state of Trivium is 288 bits and it receives a secret key of 80 bits and an IV of 80 bits. A block diagram of Trivium is depicted in Fig. 2.

3.3.1. Design Specification of Trivium

The internal state of the cipher can be regarded as three NLFSRs with input, namely NLFSR93, NLFSR84, and NLFSR111 of lengths 93, 84 and 111, respectively. We denote by $(a_i, a_{i+1}, ..., a_{i+92})$ an internal state of NLFSR93, $(b_i, b_{i+1}, ..., b_{i+83})$ an internal state of NLFSR84, and $(c_i, c_{i+1}, ..., c_{i+110})$ an internal state of NLFSR111. The recurrence relations for the internal state update and the function for the keystream $\{z_i\}$ generation of Trivium are defined

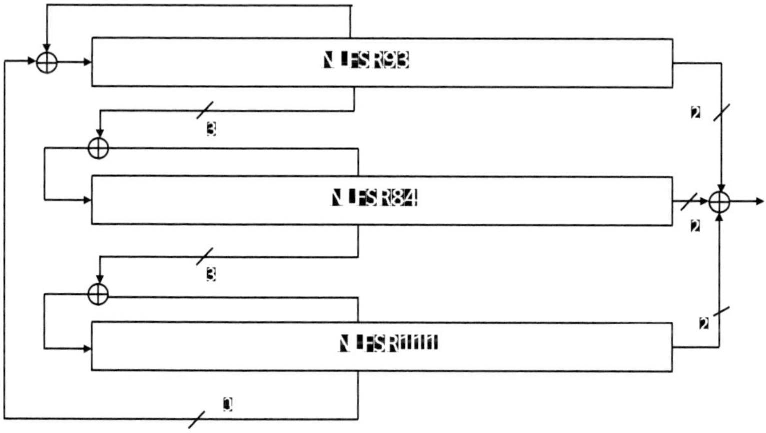

Figure 2. An architecture of Trivium.

as follows:

$$\begin{aligned} b_{i+84} &= a_i + a_{i+27} + a_{i+1}a_{i+2} + b_{i+6}, i \geq 0 \\ c_{i+110} &= b_i + b_{i+15} + b_{i+1}b_{i+2} + c_{i+24}, i \geq 0 \\ a_{i+93} &= c_i + c_{i+45} + c_{i+1}c_{i+2} + a_{i+24}, i \geq 0 \\ z_i &= a_i + a_{i+27} + b_i + b_{i+15} + c_i + c_{i+45},\ a_i, b_i, c_i, z_i \in \mathbb{F}_2, i \geq 0. \end{aligned}$$

While loading the key and IV, the key is loaded in the first 80 bit position of NLFSR94, the IV is loaded in the first 80 bit positions of NLFSR84, and NLFSR111 is initialized with 0's, except the last 3 bit positions. Before generating a keystream, the initialization phase is executed for 1152 rounds to blend the key bits and the IV bits. According to the designers, Trivium can generate a keystream of length smaller than 2^{80} with probability 2^{-208} for a key/IV pair.

3.3.2. Existing Attacks on Trivium and Implementation

Fischer, Khazaei, and Meier presented a key recovery attack on a 672-round variant of Trivium with complexity 2^{55} [30]. Dinur and Shamir [23] applied the cube attack on Trivium with 672-, 735-, and 767-round of initialization and the attack can recover the full key with 2^{19}, 2^{30}, and 2^{45} bit operations, respectively. Very recently, at FSE 2013, Fouque and Vannet [29] presented an improved key-recovery attack using the cube attacking technique for 784-round and 799-round of initialization of Trivium. The improved attack can recover 42 key bits using 42 linear equations of 784-round Trivium with less than 2^{38} cipher requests in the online phase and other 38 bits are recovered by the exhaustive search. Consequently, the 80-bit key can be recovered in less than 2^{39} cipher requests. For 799-round Trivium, the complexity of the improved attack to recover 80-bit key is 2^{62} where 12 key bits are found using 12 linear equations, 6 key bits are obtained using 6 quadratic equations, and other 62 key bits are recovered by the exhaustive search. Some attacks on Trivium and its variants can also be found in [8, 52].

3.4. Stream Cipher: WG-8

Fan, Mandal and Gong recently proposed WG-8 stream cipher, which is a lightweight variant of the WG stream cipher [28]. WG-8 was designed for providing confidentiality in resource-constrained environments. The WG-8 cipher accepts an 80-bit secret key and an 80-bit initial vector and it generates keystreams with good randomness properties such as maximum period, 2-level autocorrelation and high linear complexity. Fig. 3 illustrates an overview of the WG-8 stream cipher.

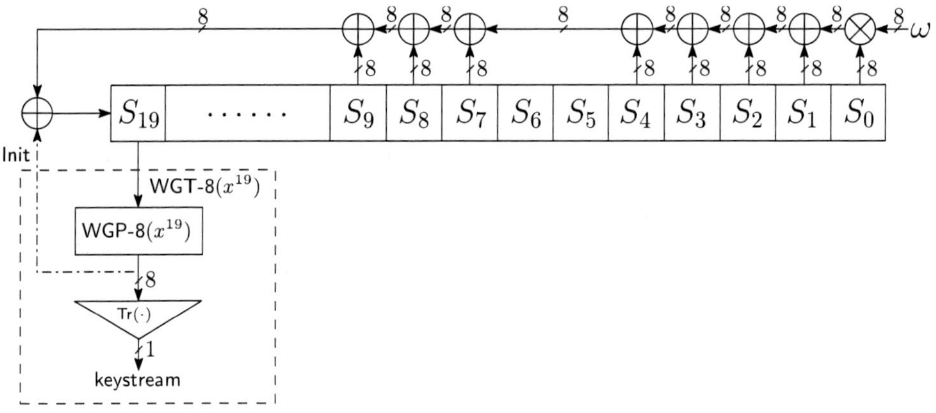

Figure 3. An architecture of the WG-8 stream cipher.

3.4.1. Design Details of WG-8 Stream Cipher

WG-8 stream cipher is composed of an LFSR of length 20 and a WG transformation over $\mathbb{F}_{2^8}$. The LFSR uses the primitive polynomial $f(x) = \omega + x +$

$x^2 + x^3 + x^4 + x^7 + x^8 + x^9 + x^{20}$ over $\mathbb{F}_{2^8}$ for generating an m-sequence of period $2^{160} - 1$ over $\mathbb{F}_{2^8}$. The WG-8 permutation, denoted by WGP-8, and WG-8 transformation, denoted by WGT-8, are defined as

$$\begin{aligned}\text{WGP-8}(x) &= x + (x+1)^{2^3+1} + (x+1)^{2^6+2^3+1} + (x+1)^{2^6-2^3+1} \\ &\quad + (x+1)^{2^6+2^3-1}, x \in \mathbb{F}_{2^8} \\ \text{WGT-8}(x) &= \text{Tr}\Big(x + (x+1)^{2^3+1} + (x+1)^{2^6+2^3+1} + (x+1)^{2^6-2^3+1} \\ &\quad + (x+1)^{2^6+2^3-1}\Big), x \in \mathbb{F}_{2^8}\end{aligned}$$

where $\text{Tr}(\cdot)$ is the trace function mapping from $\mathbb{F}_{2^8}$ to $\mathbb{F}_2$. Let $(S_0, S_1, ..., S_{19})$ be the internal state of the LFSR and $\{s_i\}$ be a keystream produced by the WG-8 cipher. The cipher is executed as follows

$$a_{k+20} = \begin{cases} \omega S_k + S_{k+1} + S_{k+2} + S_{k+3} + S_{k+4} + S_{k+7} + S_{k+8} + S_{k+9} + \text{WGP-8}(S_{k+19}^{19}), \\ S_i, \in \mathbb{F}_{2^8},\ 0 \le k < 40 \text{ (initialization phase)} \\ S_k + S_{k+1} + S_{k+2} + S_{k+3} + S_{k+4} + S_{k+7} + S_{k+8} + S_{k+9}, k \ge 40 \text{ (running phase)} \end{cases}$$
$$s_k = \text{WGT-8}(S_{40+k}^{19}),\ S_k \in \mathbb{F}_{2^8}, s_k \in \mathbb{F}_2,\ k \ge 0.$$

The decimation $d = 19$ is used to strengthen the WG transformation as it increases the linear complexity of the keystream and the algebraic degree of the WG transformation. Keystream $\{s_i\}$ has the following randomness properties: a) period $2^{160} - 1$; b) balance; c) ideal 2-level autocorrelation; d) ideal t-tuple distribution $1 \le t \le 20$; and e) linear span $2^{32.32}$. A software implementation using directly look-up table, coset leader based look-up table, and tower field arithmetic approaches can be found in [28].

3.4.2. Existing Attacks on WG-8 and Implementation

According to the designers of WG-8, the cipher is secure against algebraic attacks, cube attacks, distinguishing attacks, correlation attacks, and discrete fourier transformation (DFT) attacks [28]. The best know attack on WG-8 is the DFT attack whose time complexity is $O(2^{33.32})$ where one is required $O(2^{33.32})$ consecutive keystream bits to launch the attack.

3.5. Stream Cipher: WG-5

WG-5 stream cipher, proposed by Aagaard, Gong and Mota in [1], is a synchronous stream cipher based on the WG transformation over $\mathbb{F}_{2^5}$, which is also a lightweight variant of the WG stream cipher family. The main goal of designing the WG-5 stream cipher is to secure passive RFID applications with low-cost and low-power requirements. The design of the WG-5 cipher is similar to the design of the WG-8 stream cipher in Fig. 3. The WG-5 stream cipher uses an LFSR of length 32 and a WG transformation over $\mathbb{F}_{2^5}$. Two variants of WG-5 can be found in [1]. One variant uses the WG transformation $WG_5(x)$ and another one uses $WG_5(x^{11})$, which is a WG transformation with decimation 11 used to increase the linear span of the keystream. The cipher accepts a secret key of 80 bits and an IV of 80 bits and produces a keystream with the following randomness properties: a) period $2^{160}-1$; b) balance; c) ideal 2-level autocorrelation; d) t-tuple distribution, $1 \leq t \leq 32$; and e) linear span 2^{17} for $WG_5(x)$ and 2^{22} for $WG_5(x^{11})$.

The security of the WG-5 cipher is dependent on the security of the WG transformation used in the cipher. Similar to the WG-8 cipher, the best known attack on the WG-5 cipher is the DFT attack whose time complexity is $O(2^{17})$ for the variant uses $WG_5(x)$ and $O(2^{22})$ for the variant uses $WG_5(x^{11})$, according to the analysis of [1].

3.6. Summary of Hardware Implementations of the Stream Ciphers

This section presents the hardware implementation costs of the presented four stream ciphers. Hardware implementations of Grain stream ciphers can be found in [37, 34, 33]. An ASIC implementation of Grain-128 that outputs one bit per clock cycle requires 2133 GEs, and an implementation that outputs 32 bits per clock cycle requires 4768 GEs [34]. An FPGA implementation of Grain-128 on Spartan-3 requires 50 slices [37]. Several hardware implementations of Trivium have been published in [37, 70, 1, 18]. To the best of our knowledge, the ASIC implementation of Trivium in [70] on $0.13\ \mu$m CMOS technology requires 1986 GEs and an FPGA implementation of Trivium in [37] on Spartan-3 device requires 50 slices where Trivium outputs one bit in each clock cycle. A software implementation of the WG-8 cipher can be found in [28]. Yang et al. [70] presented an efficient hardware implementation of the WG-8 cipher for FPGAs and ASICs. An FPGA implementation shows that one requires 137 slices to implement WG-8 cipher on a Xilinx Spartan-3 FPGA platform and the cost of the ASIC implementation on 65nm CMOS technology is 1786 GEs using the look-up table where the cipher outputs one bit in each clock cycle. Two variants of the WG-5 ciphers are developed by using two different WG transformations. An ASIC implementation for the variants that use $WG_5(x)$ and $WG_5(x^{11})$ require 1350 GEs and 1360 GEs, respectively [1]. A summary of the performance comparison of four stream ciphers is provided in Table 3.

Table 3. A performance comparison for stream ciphers

Cipher	Internal state (bits)	Key (bits)	IV (bits)	Area (GEs)	Area (Slices)	Throughput Kb/s	Logic Process
Grain-128 [37, 34]	256	128	96	2133	50	196 (Mbps)	XC3S50-5PQ208
Grain-80 [37, 33]	160	80	64	1126	44	196 (Mbps)	XC3S50-5PQ208
Trivium [37, 70]	288	80	64	1986	50	240 (Mbps)	XC3S50-5PQ208
WG-8 (LUT) [70]	160	80	80	1786	137	190 (Mbps)	Spartan-3(xc3s1000)
WG-8 (TF) [70]	160	80	80	3162	343	42 (Mbps)	Spartan-3(xc3s1000)
WG-5(x) [1]	160	80	80	1350	–	200	0.13 μm
WG-5 (x) [1]	160	80	80	1508	–	200	0.18 μm
WG-5 (x^{11}) [1]	160	80	80	1360	–	200	0.13 μm
WG-5 (x^{11})[1]	160	80	80	1521	–	200	0.18 μm

4. Lightweight (Pseudo)Random Number Generators

A random number generator is an essential component for implementing security functionalities of a device. There are two types of random number generators, namely *true or hardware random number generators* (TRNGs) implemented by sampling some physical phenomena and *pseudorandom number generators* (PRNGs) designed using mathematical functions. The EPC Class 1 Generation 2 (EPC C1 Gen2 in brief) is an international ISO18000-6C standard for radio frequency identification tags. In the EPC C1 Gen2 standard, the tag identification protocol requires 2 or 3 16-bit random numbers to identify a tag and the security of the tag relies on the in-built random number generator. In this section, we present four random number generators dedicated to the EPC C1 Gen2 RFID tags.

4.1. Che et al.'s PRNG

Che et al. [20] designed a PRNG by employing a combination of an oscillator-based TRNG and a linear feedback shift register (LFSR) with 16 stages. An overview of the design is provided in Fig. 4. In their design, the TRNG is

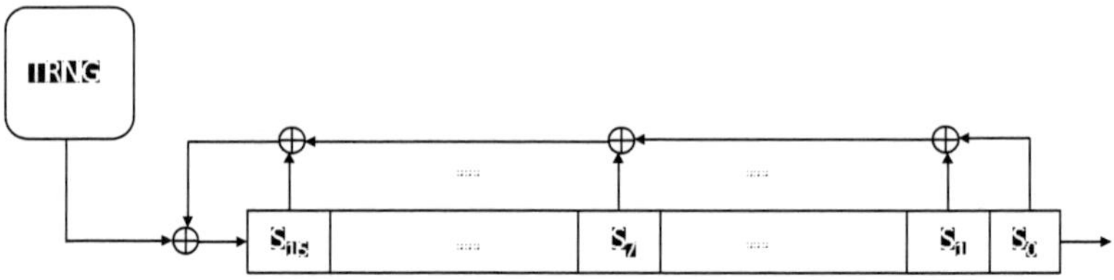

Figure 4. A block diagram of Che et al.'s PRNG.

implemented using an analog circuit and exploits thermal noise of the circuit. Considering low power consumption and introducing randomness, a true random bit is generated by the TRNG in 16 clock cycles and the truly random bit is XORed with each bit of a 16-bit sequence generated from the LFSR. Thus, in 16 clock cycles, a 16-bit random number is generated by the random number generator. Due to the linear structure in the design, Che et al.'s scheme has been attacked by Melia-Segui et al. in [53] with a high success probability $\frac{(n+1)}{8n}$, where n is the length of the LFSR.

4.2. Melia-Segui et al.'s PRNG

Melia-Segui et al. [53] proposed a similar design by employing multiple primitive polynomials instead of one in the LFSR of length 16. The reason for using an LFSR with multiple primitive polynomials is to prevent a similar attack applied on Che et al.'s PRNG. The design of their random number generator consists of a true random source, a module with eight primitive polynomials, and a decoding circuit taking inputs from the true random source, where the decoding circuit is designed in such a way that the same primitive polynomial is not chosen consecutively. An overview of the design is given in Fig. 5.

At each clock cycle, one primitive polynomial is chosen according to the decoding logic and true random bits for producing a pseudorandom bit. Thus, the PRNG produces a 16-bit random number in 16 clock cycles. Note that to produce one bit from the LFSR, three true random bits are required. The security of the PRNG completely relies on the TRNG, as the length of the LFSR is short.

Recently, Melia-Segui et al. [54] proposed J3Gen which contains four instances of PRNGs for different lengths of the LFSR with different number of primitive polynomials. The design principle of J3Gen is also based on an LFSR with multiple primitive polynomials. The security properties of all PRNGs were

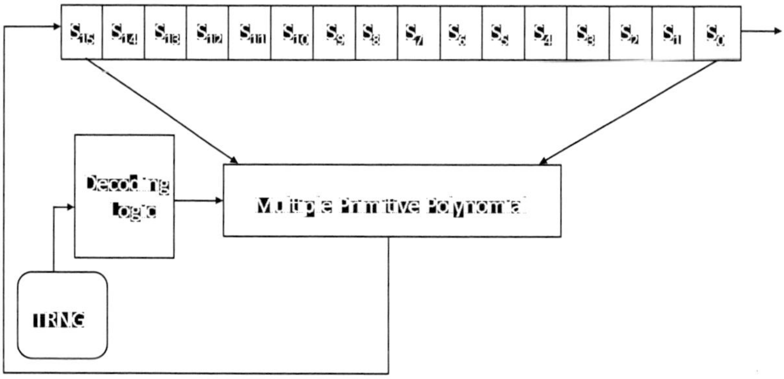

Figure 5. An overview of Melia-Segui et al.'s PRNG.

analyzed by performing the statistical tests proposed by the EPC C1 Gen2 standard.

4.3. PRNG: LAMED

In [60], Peris-Lopez et al. proposed a PRNG named LAMED for RFID tags, which is in compliance with the EPC C1 Gen2 standard and can provide 32-bit as well as 16-bit random numbers. The basic operations for updating the internal state of LAMED consist of bitwise XOR operations, modular algebra, and bit rotations. The internal state of the LAMED is of 64-bit, including a 32-bit key and a 32-bit initial vector. The key length can be further increased by replacing the IV bits with the key bits. Note that the LAMED always outputs a 32-bit random number and a 16 bit random number is obtained by dividing 32-bit number into two equal halves and XORing them together.

4.4. PRNG: Warbler

Warbler, recently designed by Mandal, Fan and Gong in [51], is a lightweight pseudorandom number generator for the EPC C1 Gen2 RFID tags for generating 16-bit random numbers. Warbler is a purely NLFSR-based PRNG, which is designed to provide guaranteed randomness properties and to prevent the existing cryptanalytic attacks, and to have an efficient hardware implementation. Warbler is composed of two building blocks, the first building block contains two NLFSRs that generate span n sequences, and the second building block is a nonlinear WG generator based on WG transformations. The length of the internal state of Warbler is 65 bits, including 45 bits for the key and 20 bits for the IV. The architecture of Warbler is provided in Fig. 6.

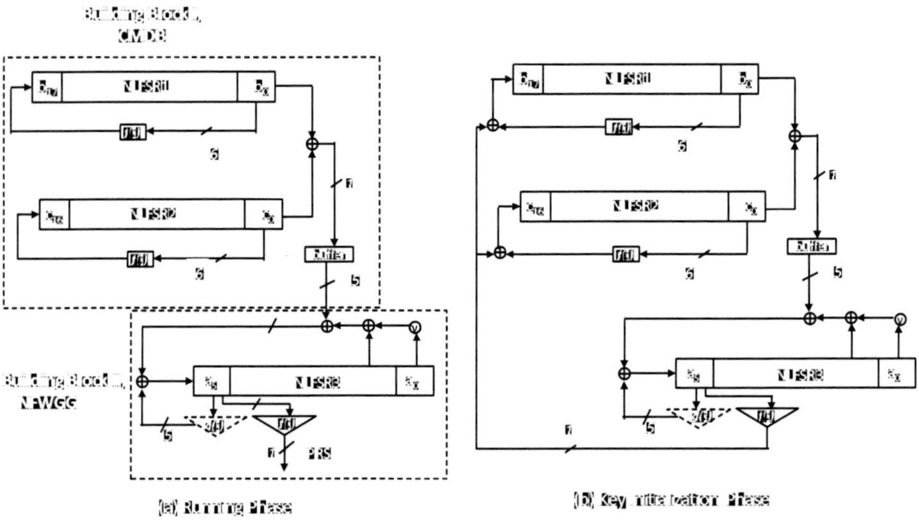

Figure 6. A block diagram of the Warbler PRNG for EPC C1 Gen2 tags.

4.4.1. Design Specification of Warbler

The first building block of Warbler contains two NLFSRs, namely NLFSR18 and NLFSR17 of lengths 17 and 18, and both NLFSRs generate span n sequences of period $2^{18} - 1$ and $2^{17} - 1$, respectively. The finite field $\mathbb{F}_{2^5}$ is defined by α and $\alpha^5 + \alpha^4 + \alpha^3 + \alpha + 1 = 0$. We denote by $(b_i, b_{i+1}, ..., b_{17+i})$ and $(c_i, c_{i+1}, ..., c_{16+i})$ the internal states of NLFSR18 and NLFSR17, respectively. The internal states of NLFSR18 and NLFSR17 are updated as

$$b_{i+18} = b_i + f(b_{i+4}, b_{i+7}, b_{i+8}, b_{i+10}, b_{i+15}), b_i \in \mathbb{F}_2, i \geq 0$$
$$c_{i+17} = c_i + f(c_{i+4}, c_{i+7}, c_{i+8}, c_{i+9}, c_{i+12}), c_i \in \mathbb{F}_2, i \geq 0$$

where $f(x_1, x_2, \cdots, x_5) = x_1 + x_4 + x_1x_2 + x_1x_3 + x_1x_4 + x_1x_5 + x_2x_3 + x_2x_5 + x_3x_5 + x_4x_5 + x_1x_2x_4 + x_1x_2x_5 + x_1x_3x_4 + x_1x_4x_5 + x_2x_3x_5$ is a WG transformation over $\mathbb{F}_{2^5}$. A binary sequence $\mathbf{s} = \{s_i\}$ is obtained as $s_i = b_i + c_i$ and the sequence $\mathbf{s}$ has the following randomness properties: a) period $(2^{18}-1)(2^{17}-1) \approx 2^{35}$; and b) linear complexity $2^{18.58}$. A new sequence $\mathbf{t} = \{t_i\}$ over $\mathbb{F}_{2^m}$ is obtained from the sequence $\mathbf{s}$ as follows

$$t_i = (s_{5i}, s_{5i+1}, s_{5i+2}, s_{5i+3}, s_{5i+4}), i \geq 0.$$

The sequence $\mathbf{t}$ is fed in the second building block to bound the period and linear complexity of the output sequence.

The recurrence relation for the internal state update for the nonlinear WG generator is given by

$$a_{i+6} = \gamma a_i + a_{i+1} + t_i + w_i, \quad \gamma, t_i, a_i, w_i = (0, ..., g(a_{i+5})) \in \mathbb{F}_{2^5}, i \geq 0$$

where $g(x_1, x_2, \ldots, x_5) = x_1 + x_1x_4 + x_2x_3 + x_3x_5 + x_1x_2x_3 + x_1x_2x_5 + x_1x_3x_5 + x_1x_4x_5 + x_2x_3x_5 + x_2x_4x_5 + x_3x_4x_5$ is a WG transformation and $\gamma = \alpha^{15}$. The output sequence $\{o_i\}$ is obtained as $o_i = f(a_{k+5})$ and 16-bit random numbers $\{R_i\}$ are obtained as

$$R_i = (o_{16i}, o_{16i+1}, o_{16i+2}, ..., o_{16i+15}).$$

The period and linear complexity of the output sequence $\{o_i\}$ are bounded below by $2^{37.32}$ and $2^{18.58}$, respectively. The WG transformation f is used as a filtering function with nonlinearity 12, algebraic degree 3 and algebraic immunity 3. Before generating random numbers, an initialization phase of 44 rounds is applied. For the details of the initialization phase and cryptanalysis, we refer the readers to [51]. A high level overview of the initialization phase of Warbler can be found in Fig. 6 (b). For a general architecture and the randomness properties of the Warbler family, we refer the readers to Mandal's PhD thesis [50].

4.5. A Summary of Hardware Implementations of the PRNGs

We provide a comparison of the hardware implementation of presented random number generators. Since Melia-Segui et al.'s PRNG is composed of a TRNG and an LFSR of length 16, the hardware cost of the LFSR component is 761 GEs while the cost for the TRNG component is not included. The designers of LAMED estimated that the PRNG can be implemented using 1585 GEs and the implementation requires 1.8 ms to generate a 32-bit random number [60]. Warbler was proposed to reduce the hardware implementation cost and increase throughput and high security. Warbler can be implemented on low-cost Xilinx Spartan-3 FPGA devices with 46 slices.

Table 4. A performance comparison for PRNGs

PRNG	Internal state (bits)	Key (bits)	IV (bits)	Area (GEs/Slices)	Randomness Period	Randomness Linear span
Che et al. [20]	16	–	–	–	–	–
Melia-Segui et al. [53]	16	–	–	761 GEs	–	–
LAMED [60]	64	32	32	1585 GEs	–	–
Warbler [51]	65	45	20	46 Slices/ 760 GE	$\geq 2^{37.32}$	$\geq 2^{18.58}$

Conclusion

In this chapter, we present a research survey on the lightweight symmetric-key cryptographic primitives including block ciphers, stream ciphers, and pseudo-random number generators that are dedicatedly designed for securing passive and low-cost RFID systems. Lightweight cryptography is able to provide adequate security for protecting resource constrained smart devices like RFID tags without introducing significant overhead and cost. A research survey about the innovative ideas in those lightweight cryptographic primitives will further stimulate new designs in this field and provide valuable guidance for building a secure and cost-effective RFID system in practice.

References

[1] M.D. Aagaard, G. Gong and R.K. Mota, Hardware Implementations of the WG-5 Cipher for Passive RFID Tags. *2013 IEEE International Symposium on Hardware-Oriented Security and Trust (HOST)*, pp. 29 – 34, 2013.

[2] M.A. Abdelraheem, G. Leander, and E. Zenner, Differential Cryptanalysis of Round-Reduced PRINTcipher: Computing Roots of Permutations. *Fast Software Encryption*, LNCS, vol. 6733, pp. 1 – 17, Springer, 2011.

[3] M.A. Abdelraheem, G. Leander, H. AlKhzaimi and E. Zenner, A Cryptanalysis of PRINTcipher: The Invariant Subspace Attack. *Advances in Cryptology – CRYPTO 2011*, LNCS, vol. 6841, pp. 206 – 221, Springer, 2011.

[4] F. Abed, E. List, S. Lucks, and J. Wenzel, *Differential and Linear Cryptanalysis of Reduced-Round Simon*. Cryptology ePrint Archive, Report 2013/526, 2013. http://eprint.iacr.org/

[5] Z. Ahmadian, M. Salmasizadeh, and M.R. Aref, *Biclique Cryptanalysis of the Full-Round KLEIN Block Cipher*. Cryptology ePrint Archive, Report 2013/097, 2013. http://eprint.iacr.org/

[6] H.A. Alkhzaimi and M.M. Lauridsen, *Cryptanalysis of the SIMON Family of Block Ciphers*. Cryptology ePrint Archive, Report 2013/543, 2013. http://eprint.iacr.org/

[7] J.-P. Aumasson, I. Dinur, L. Henzen, W. Meier, and A. Shamir, Efficient FPGA Implementations of High-Dimensional Cube Testers on the Stream Cipher Grain-128. In *Special-purpose Hardware for Attacking Cryptographic Systems (SHARCS)*, 2009.

[8] Aumasson, Jean-Philippe and Dinur, Itai and Meier, Willi and Shamir, Adi, Cube Testers and Key Recovery Attacks on Reduced-Round MD6 and Trivium. In *Fast Software Encryption (FSE 2009)*, LNCS, vol. 5665, pp. 1 – 22, Springer, Heidelberg, 2009.

[9] J.-P. Aumasson, M. Naya-Plasencia, and M.-J. Saarinen, Practical Attack on 8 Rounds of the Lightweight Block Cipher KLEIN. In *Progress in Cryptology C INDOCRYPT 2011*, LNCS, vol. 7107, pp. 134 – 145, Springer, 2011.

[10] X. Bai, L. Huang, Y. Wang, and Y. Xu, Differential Power Analysis Attack on CLEFIA Block Cipher. In *Computational Intelligence and Software Engineering (CiSE 2009)* pp. 1 – 4, 2009.

[11] R. Beaulieu, D. Shors, J. Smith, S. Treatman-Clark, B. Weeks, and L. Wingers, *The SIMON and SPECK Families of Lightweight Block Ciphers*. Cryptology ePrint Archive, Report 2013/404, 2013. http://eprint.iacr.org/

[12] A. Bogdanov, L.R. Knudsen, G. Leander, C. Paar, A. Poschmann, M.J.B. Robshaw, Y. Seurin, C.H. Vikkelsoe, PRESENT: An Ultra-Lightweight Block Cipher. *Proceedings of Cryptographic Hardware and Embedded Systems (CHES)*, LNCS, vol. 4727, pp. 450 – 466, Springer, 2007.

[13] A. Bogdanov, and C. Rechberger, A 3-Subset Meet-in-the-Middle Attack: Cryptanalysis of the Lightweight Block Cipher KTANTAN. In *Selected Areas in Cryptography (SAC 2010)*, LNCS, vol. 6544, pp. 229 – 240, 2010.

[14] J. Borghoff, A. Canteaut, T. Güneysu, E.B. Kavun, M. Knezevic, L.R. Knudsen, G. Leander, V. Nikov, C. Paar, C. Rechberger, P. Rombouts, S.S. Thomsen, and T. Yalcin, PRINCE - A Low-Latency Block Cipher for Pervasive Computing Applications - Extended Abstract, *ASIACRYPT 2012*, LNCS, vol. 7658, pp. 208 – 225, Springer, 2012.

[15] A. Biryukov, and I. Nikolić, *Security Analysis of the Block Cipher CLEFIA*, http://www.cryptrec.go.jp/estimation/techrep_id2202-2.pdf.

[16] C. De Canníere, O. Dunkelman, and M. Knezevic, KATAN and KTANTAN – A Family of Small and Efficient Hardware-Oriented Block Ciphers. In *Cryptographic Hardware and Embedded Systems - CHES 2009*, LNCS, vol. 5747, pp. 272 – 288, Springer, 2009.

[17] C. De Canniére, and B. Preneel, *TRIVIUM – A Stream Cipher Construction Inspired by Block Cipher Design Principles*. http://www.ecrypt.eu.org/stream/trivium.html

[18] C. De Canniére, and B. Preneel, *TRIVIUM Specifications*. http://www.ecrypt.eu.org/stream/ciphers/trivium/trivium.pdf

[19] C. De Canniére, O. Kücük, and B. Preneel, Analysis of Grain's Initialization Algorithm. In *SASC 2008*, 2008.

[20] W. Che, H. Deng, W. Tan, and J. Wang, A Random Number Generator for Application in RFID Tags. In *Networked RFID Systems and Lightweight Cryptography*, pp. 279 – 287, Springer-Verlag, 2008.

[21] L. Chen, and G. Gong, *Communication System Security*, Boca Raton, Florida, USA: Chapman & Hall/CRC, 2012.

[22] B. Collard, and F.-X. Standaert, A Statistical Saturation Attack against the Block Cipher PRESENT. In Fischlin (eds.) *CT-RSA 2009*, LNCS, vol. 5473, pp. 195 – 210, Springer, 2009.

[23] I. Dinur, and A. Shamir, Cube Attacks on Tweakable Black Box Polynomials. In *Advances in Cryptology - EUROCRYPT 2009*, LNCS, vol. 5479, pp. 278 – 299, 2009.

[24] I. Dinur, and A. Shamir, Breaking Grain-128 with Dynamic Cube Attacks. In *Fast Software Encryption (FSE 2011)*, LNCS, vol. 6733, pp. 167 – 187, 2011.

[25] T. Eisenbarth, S. Kumar, C. Paar, A. Poschmann, and L. Uhsadel, A Survey of Lightweight-Cryptography Implementations. *IEEE Des. Test*, vol. 24, No. 6, pp. 522 – 533, 2007.

[26] H. Englund, T. Johansson, and M.S. Turan, A Framework for Chosen IV Statistical Analysis of Stream Ciphers. In K. Srinathan, C. Pandu Rangan, and Moti Yung (eds.), *INDOCRYPT 2007*, LNCS, vol. 4859, pp. 268 – 281, Springer, 2007.

[27] *The eStream Project.* http://www.ecrypt.eu.org/stream/

[28] X. Fan, K. Mandal, and G. Gong, WG-8: A Lightweight Stream Cipher for Resource-Constrained Smart Devices. In K. Singh, A.K. Awasthi, and R. Mishra (Eds.): *QSHINE 2013*, LNICST 115, pp. 617 – 632, 2013.

[29] P.A. Fouque and T. Vannet, Improving Key Recovery to 784 and 799 rounds of Trivium using Optimized Cube Attacks. In *FSE 2013*, Springer-Verlag, 2013.

[30] S. Fischer, S. Khazaei, and W. Meier, Chosen IV Statistical Analysis for Key Recovery Attacks on Stream Ciphers. In *AFRICACRYPT 2008*, LNCS, vol. 5023, pp. 236 – 245, 2008.

[31] Z. Gong, S. Nikova, and Y.W. Law, KLEIN: A New Family of Lightweight Block Ciphers. In *RFID Security and Privacy - RFIDSec 2011*, LNCS, vol. 7055, pp. 1 – 18, Springer, 2011.

[32] D.H. Green and K.R. Dimond, Nonlinear Product-Feedback Shift Registers. *Proceeding IEE 117*, pp. 681 – 686, 1970.

[33] M. Hell, T. Johansson, and W. Meier, Grain: A Stream Cipher for Constrained Environments. *Int. J. Wire. Mob. Comput.*, vol. 2, pp. 86 – 93, 2007.

[34] M. Hell, T. Johansson, A. Maximov, and W. Meier, A Stream Cipher Proposal: Grain-128. *2006 IEEE International Symposium on Information Theory*, pp. 1614 – 1618, 2006.

[35] D. Hong, J. Sung, S. Hong, J. Lim, S. Lee, B.-S. Koo, C. Lee, D. Chang, J. Lee, K. Jeong, H. Kim, J. Kim, and S. Chee, HIGHT: A New Block Cipher Suitable for Low-Resource Device. *Proceedings of Cryptographic Hardware and Embedded Systems (CHES) 2006*, LNCS, vol. 4249, pp. 46 – 59, Springer, 2006.

[36] H. Hu and G. Gong, Periods on Two Kinds of Nonlinear Feedback Shift Registers with Time Varying Feedback Functions. *International Journal of Foundations of Computer Science*, vol. 22, No. 6, pp. 1317 – 1329, 2011.

[37] D. Hwang, M. Chaney, S. Karanam, N. Ton, and K. Gaj, Comparison of FPGA-targeted Hardware Implementations of eStream Stream Cipher Candidates. *The State of the Art of Stream Ciphers*, pp. 151 – 162, 2008.

[38] ISO/IEC 29192-2:2012. *Information technology – Security techniques – Lightweight cryptography – Part 2: Block ciphers*, International Standards Organization, 2012.

[39] J. Jean, I. Nikolić, T. Peyrin, L. Wang, and S. Wu, Security Analysis of PRINCE. In *Fast Software Encryption (FSE 2013)*, LNCS, Springer, 2013.

[40] A. Juels and S. A. Weis, Authenticating Pervasive Devices with Human Protocols. In *Advances in Cryptology - CRYPTO'05*, LNCS, vol. 3621, pp. 293 – 308, Springer, 2005.

[41] P. Kitsos and Y. Zhang, RFID Security: Techniques, Protocols and System-On-Chip Design. *Springer Science+Business Media*, LLC, New York, , USA, 2008.

[42] L. Knudsen, G. Leander, A. Poschmann, and M.J.B. Robshaw, PRINTcipher: A Block Cipher for IC-Printing. In *CHES 2010*, LNCS, vol. 6225, pp. 16 – 32, Springer, 2010.

[43] S. Knellwolf, W. Meier and M. Naya-Plasencia, Conditional Differential Cryptanalysis of NLFSR-based Cryptosystems. In *Advances in Cryptology - ASIACRYPT 2010*, LNCS, vol. 6477, pp. 130 – 145, Springer, 2010.

[44] G. Leander, A. Poschmann, C. Paar, and K. Schramm, New Lightweight DES Variants. In *Fast Software Encryption (FSE 2007)*, LNCS, vol. 4593, pp. 196 – 210, Springer, 2007.

[45] Y. Li, W. Wu, and L. Zhang, Improved Integral Attacks on Reduced-Round CLEFIA Block Cipher. In *Information Security Applications*, LNCS, vol. 7115, pp. 28-39, 2012.

[46] C.H. Lim, A revised version of CRYPTON: CRYPTON v1.0. In *Fast Software Encryption (FSE 1999)*, LNCS, vol. 1636, pp.31-45, Spinger-Verlag, 1999.

[47] C.H. Lim, T. Korkishko, mCrypton – A Lightweight Block Cipher for Security of Low-Cost RFID Tags and Sensors, *Proceedings of Workshop on Information Security Applications (WISA)*, LNCS, vol. 3786, pp. 243 – 258, Springer, 2005.

[48] H. Mala, M. Dakhilalian, and M. Shakiba, Impossible Differential Attacks on 13-Round CLEFIA-128. *Journal of Computer Science and Technology*, vol. 26, No. 4, pp. 744 – 750, Springer US, 2011.

[49] H. Mala, M. Dakhilalian, and M. Shakiba, Cryptanalysis of mCrypton – A lightweight block cipher for security of RFID tags and Sensors. *International Journal of Communication Systems*, vol. 25, Issue 4, pp. 415 – 426, 2011.

[50] K. Mandal, Design and Analysis of Cryptographic Pseudorandom Number/Sequence Generators with Applications in RFID, *PhD Thesis*, University of Waterloo, 2013.

[51] K. Mandal, X. Fan, and G. Gong, Warbler: A Lightweight Pseudorandom Number Generator for EPC C1 Gen2 RFID Tags. In N.W. Lo and Y. Li (Eds.), *Cryptology and Information Security Series - The 2012 Workshop on RFID and IoT Security (RFIDsec'12 Asia)*, vol. 8, pp. 73-84, IOS Press, Amsterdam, Netherlands, 2012.

[52] A. Maximov, and A. Biryukov, Two Trivial Attacks on Trivium. In *Selected Areas in Cryptography (SAC 2007)*, LNCS, vol. 4876, pp. pp 36 – 55, Springer Heidelberg, 2007.

[53] J. Melia-Segui, J. Garcia-Alfaro, and J. Herrera-Joancomarti, Analysis and Improvement of a Pseudorandom Number Generator for EPC Gen2 Tags. In *Proceedings of the 14th International conference on Financial Cryptography and Data Security, FC'10*, pp. 34 – 46, Springer-Verlag, 2010.

[54] J. Meliá-Seguí , J. Garcia-Alfaro, and J. Herrera-Joancomartí, J3Gen: A PRNG for Low-Cost Passive RFID. *Sensors*, vol. 13, No. 3, pp. 3816 – 3830, 2013.

[55] M.J. Mihaljević, S. Gangopadhyay, G. Paul, and H. Imai, Internal State Recovery of Grain-v1 Employing Normality Order of the Filter Function, *IET Information Security*, vol. 6, No.2, pp. 55 – 64, June 2012.

[56] A. Moradi, A. Poschmann, S. Ling, C. Paar, and H. Wang. Pushing the Limits: A Very Compact and a Threshold Implementation of AES. In *Advances in Cryptology - EUROCRYPT 2011*, LNCS, vol. 6632, pp. 69 – 88, Springer, 2011.

[57] J. Nakahara Jr., P. Sepehrdad, B. Zhang, and M. Wang, Linear (Hull) and Algebraic Cryptanalysis of the Block Cipher PRESENT. In *Cryptology and Network Security*, LNCS, vol. 5888, pp. 58 – 75, Springer, 2009.

[58] O. Özen, K. Varici, C. Tezcan, and C. Kocair, Lightweight Block Ciphers Revisited: Cryptanalysis of Reduced Round PRESENT and HIGHT. In *Information Security and Privacy*, LNCS, vol. 5594, pp. 90 – 107, 2009.

[59] J.H. Park, Security Analysis of mCrypton Proper to Low-Cost Ubiquitous Computing Devices and Applications. *International Journal of Communication Systems*, vol. 22, Issue 8, pp. 959 – 969, August 2009.

[60] P. Peris-Lopez, J. Hernandez-Castro, J.M. Estevez-Tapiador, and A. Ribagorda, LAMED – A PRNG for EPC Class-1 Generation-2 RFID Specification, *Computer Standards and Interfaces*, pp. 88 – 97, January 2009.

[61] A. Poschmann, Lightweight Cryptography – Cryptographic Engineering for a Pervasive World, *Ph.D. Thesis*, Department of Electrical Engineering and Information Sciences, Ruhr-Universitäet Bochum, Bochum, Germany, 2009.

[62] C. Rebeiro and D. Mukhopadhyay, *Differential Cache Trace Attack Against CLEFIA*. Cryptology ePrint Archive, Report 2010/012, 2010. http://eprint.iacr.org/2010/012.pdf

[63] P. Rogaway, *The security of DESX*. CryptoBytes, RSA Laboratories, summer 1996.

[64] C. Rolfes, A. Poschmann, G. Leander, and C. Paar, Ultra-Lightweight Implementations for Smart Devices – Security for 1000 Gate Equivalents. *Proceedings of Smart Card Research and Advanced Applications (CARDIS)*, LNCS, vol. 5189, pp. 89 – 103, Springer, 2008.

[65] M. Shakiba, M. Dakhilalian, and H. Mala, *Non-isomorphic Biclique Cryptanalysis and Its Application to Full-Round mCrypton*. Cryptology ePrint Archive, Report 2013/141, 2013. http://eprint.iacr.org

[66] T. Shirai, K. Shibutani, T. Akishita, S. Moriai, and T. Iwata, The 128-Bit Blockcipher CLEFIA (Extended Abstract). In *Fast Software Encryption (FSE 2007)* LNCS, vol. 4593, pp. 181 – 195, 2007.

[67] P. Stankovski, Greedy Distinguishers and Nonrandomness Detectors. In *INDOCRYPT 2010*, LNCS, vol. 6498, pp. 210 – 226, Springer, 2010.

[68] M. Wang, Differential Cryptanalysis of Reduced-Round PRESENT. In: Vaudenay, S. (ed.) *AFRICACRYPT 2008*, LNCS, vol. 5023, pp. 40 – 49 Springer, Heidelberg, 2008.

[69] L. Wei, C. Rechberger, J. Guo, H. Wu, H. Wang, S. Ling, *Improved meet-in-the-middle cryptanalysis of KATANTAN* Cryptology ePrint Archive, Report 2011/201, 2011. http://eprint.iacr.org/

[70] G. Yang, X. Fan, M. Aagaard, and G. Gong, Design Space Exploration of the Lightweight Stream Cipher WG-8 for FPGAs and ASICs. *Proceedings of the Workshop on Embedded Systems Security*, pp. 1 – 10, 2013.

[71] X. Yu, W. Wu,Y. Li, L. Zhang, Cryptanalysis of Reduced-Round KLEIN Block Cipher. In: *Inscrypt 2011*, LNCS, vol. 7537, pp. 237 – 250, Springer, Heidelberg, 2012.

[72] B. Zhu and G. Gong, *Multidimensional Meet-in-the-Middle Attack and Its Applications to KATAN32/48/64*. Cryptology ePrint Archive, Report 2011/619, 2011. http://eprint.iacr.org

[73] http://tools.ietf.org/html/draft-katagi-clefia-03

In: Radio-Frequency Identification (RFID) ISBN: 978-1-63117-750-7
Editor: Xinxin Fan, pp. 153-185

Chapter 7

LOW-COMPLEXITY ATTACKS ON SELECTED LIGHTWEIGHT STREAM CIPHERS

***Qi Chai*[1,*] *and Xinxin Fan*[2,†]**
[1]Google Inc.
[2]Department of Electrical and Computer Engineering
University of Waterloo, ON, Canada

Abstract

Although Radio Frequency Identification (RFID) technology provides many attractive and unique features, the constraints in computation and storage pose a new challenge for implementing cryptographic primitives and deploying security mechanisms. While many designs for lightweight block ciphers have been proposed in the literature, lightweight stream ciphers (favored by actual practitioners) have received much less attention to date: there are few proposals and implementations and many of these are broken. In this chapter, we review four lightweight stream ciphers designed and employed in the industry: Crypto-1, Hitag-2, the Atmel Cipher, and A2U2. For Crypto-1, Hitag-2, and the Atmel Cipher, we critically analyze all known attacks in various scenarios and put them into a unified perspective. For A2U2, we report a novel key recovery attack that fully recovers the secret key by only querying the encryption twice on the victim tag and computing for approximately 0.16 seconds on a standard PC.

*E-mail address: raullenchai@google.com
†E-mail address: x5fan@uwaterloo.ca

Keywords: lightweight cryptography, stream cipher, cryptanalysis

1. Introduction

Radio Frequency Identification (RFID) enabling automatic remote identification of objects is one of the most promising technologies in the field of ubiquitous computing. Despite many attractive and exclusive characteristics of RFID technology, the constraints in computation and storage combined with the extremely low manufacturing cost of RFID tags pose a new challenge that goes beyond the traditional cipher design paradigms – potentially stimulating new designs for lightweight symmetric ciphers. Though a considerable number of designs for lightweight block ciphers have been proposed, i.e., PRESENT [4], KATAN [13], LED [24], Piccolo [35], PRINCE [2], FIDES [1], SPECK & SIMON [8], lightweight stream ciphers have received much less attention to date except for a handful of proposals involving Crypto-1 [30], Hitag-2 [39], the Atmel Cipher [26], A2U2 [17], WG-7 [27], and WG-8 [19]. However, lightweight stream ciphers are actually favored by practitioners, and are widely employed in numerous real-world applications thanks to their small footprint and high performance on various hardware platforms.

1.1. Our Contributions

In this chapter, we review four lightweight stream ciphers designed and used in the industry: Crypto-1, Hitag-2, the Atmel Cipher, and A2U2. For Crypto-1, Hitag-2, and the Atmel Cipher, we analyze all known attacks with various assumptions and put them into a unified perspective. Additionally, we report a novel key recovery attack for A2U2, which is the most lightweight cryptographic primitive [17] proposed so far. Our attack can fully recover the secret key by only querying the encryption twice on the victim tag and solving 32 sparse systems of linear equations in the worst case – which takes only about 0.16 seconds on a standard PC. The goal of this chapter is to demonstrate the extent to which the lightweight stream ciphers that are proposed or currently in use are broken, and to encourage new designs. We also identify some of the common pitfalls in designing lightweight stream ciphers, in order to improve the quality of future proposals.

1.2. Organization

This chapter is organized as follows. In Section 2, we focus on the security of the Crypto-1 cipher, which is employed by MiFare Classic – the most widely-used contact-less smart card. In Section 3, the Hitag-2 – successor of Crypto-1 – is presented, and its design flaws and several attacks are described. In Section 4, we review the design and vulnerabilities of the Atmel Cipher, a recent proprietary stream cipher that has been used in the Atmel SecureMemory, CryptoMemory and CryptoRF chip families. The ultra-efficient key recovery attack against A2U2 is presented in Section 5 together with experimental results. We conclude this chapter in Section 6 with the lessons learned.

1.3. Notation

Unless otherwise stated, we employ the following notations throughout this chapter.

- P stands for the plaintext, C stands for the ciphertext, K stands for the secret key, and $ks_0...ks_{n-1}$ for the resulting n-bit key stream produced.
- $+$ denotes the XOR operation, $\odot$ the NAND operation, and $\cdot$ the AND operation.

2. Crypto-1

MiFare Classic is the most widely-used contact-less smart card – it has, for example, been used in London's Oyster card, the Netherlands' OV-Chipcard, Boston's CharlieCard, and in numerous other wireless access control and ticketing systems worldwide. Its design and implementation details were kept secret by its manufacturer until Karsten et al. [30] reverse engineered the chip in an automated way. Three cryptographic primitives embedded in MiFare Classic chips were subsequently revealed: a lightweight stream cipher with a 48-bit key, called Crypto-1; the corresponding authentication protocol; and the pseudorandom number generator used in the authentication protocol. In this section, we concentrate on the security of this cipher.

2.1. Specification

The Crypto-1 cipher, shown below, is based on nonlinearly filtering a sequence produced by a 48-bit linear feedback shift register (LFSR) that is defined by $g(x) = x^{48} + x^{43} + x^{39} + x^{38} + x^{36} + x^{34} + x^{33} + x^{31} + x^{29} + x^{24} + x^{23} + x^{21} + x^{19} + x^{13} + x^9 + x^7 + x^6 + x^5 + 1$, where "+" is the addition over $\mathbb{F}_2^{48}$ (a 48-dimensional vector space over the finite field with 2 elements). The nonlinear Boolean functions, f_a, f_b, f_c, are defined by their truth tables respectively, e.g., $f_a = 0\text{x}2c79$ indicates that $f_a(i), 0 \leq i \leq 15$, outputs the ith bit of $0\text{x}2c79$.

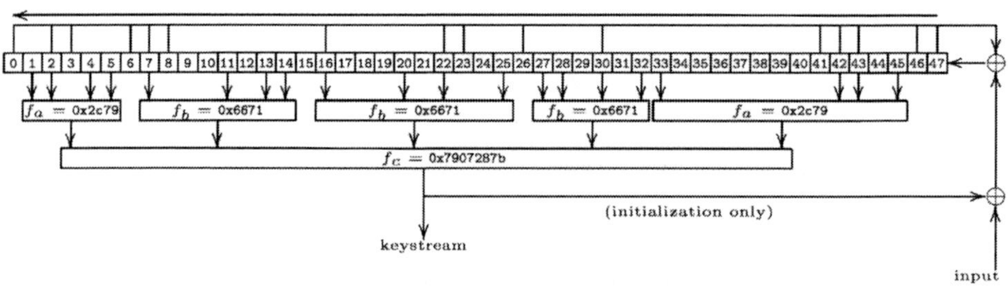

Figure 1. Stream Cipher Crypto-1.

Initialization

During initialization, the 48-bit secret key is loaded into the LFSR. At every clock tick the register is shifted one bit to the left; the leftmost bit is discarded and the feedback bit is computed according to $g(x)$. Additionally, one bit of the IV is XOR-ed with the LFSR's feedback bit and fed into the LFSR on the right. When this cipher is used in the authentication, the IV is the XOR of the tag's identity and a random number chosen by the tag.

Key Stream Generation

To produce a key stream bit at a clock tick, 20 selected bits of the LFSR are put through a nonlinear filtering function:

$$\begin{aligned} f(x) &= f_c(f_a(x_1, x_2, x_4, x_5), f_b(x_7, x_{11}, x_{13}, x_{14}), f_b(x_{16}, x_{22}, x_{24}, x_{25}), \\ &\quad f_b(x_{27}, x_{28}, x_{30}, x_{32}), f_a(x_{33}, x_{42}, x_{43}, x_{45})), \end{aligned}$$

where $(x_0 ... x_{47})$ is the internal state of the LFSR. After outputting this bit, the LFSR updates its state accordingly. This process is repeated until sufficient key stream bits are produced.

Mutual Authentication Protocol

The three-pass mutual authentication protocol employed in MiFare Classic tags is:

1. After the 32-bit nonce n_T is sent by the tag, both the tag and reader initialize the cipher with the shared key K, shift in $(uid + n_T)$, and produce the 32-bit key stream ks_1. The tag's nonce n_T is generated by the on-tag pseudorandom number generator, which is instantiated by a 16-bit LFSR.

2. The reader then picks its challenge nonce n_R and encrypts it using ks_1. Next, the reader shifts in n_R to the cipher's state to produce another key stream ks_2 of 32 bits and computes $(L^{64}(n_T) + ks_2)$, where $L^i(x)$ computes the ith state of the LFSR starting from state x. Finally the reader sends back $(n_R + ks_1)$ and $(L^{64}(n_T) + ks_2)$ to authenticate itself.

3. Upon receiving the reader's response, the tag updates the cipher state using n_R as well, and produces ks_2 and ks_3. It uses ks_2 to decrypt $(L^{64}(n_T) + ks_2)$, and if the decrypted stream is valid it responds with $(L^{96}(n_T) + ks_3)$ to finish the authentication protocol.

4. If the tag does not respond in Step 3, the reader sends $(halt + ks_3)$ to terminate the session, where $halt$ is a constant.

2.2. Cryptanalytic Attacks

Unfortunately, all of these primitives are seriously flawed, i.e., the 48-bit key of Crypto-1 can be exhaustively searched within 50 minutes using 64 FPGAs [30], and the random numbers generated are highly predictable because the 16-bit LFSR has period 65535 and repeats its state after about 618 ms.

Statistical Attack

Nohl et al. announced the first cryptographic attack against Crypto 1 [29]. This attack exploits the statistical weaknesses of the cipher in combination with the weakness of the on-tag pseudorandom number generator described above. The key observation here is that the filtering functions have poor resiliencies, such that their output bits are significantly biased if one bit is held constant. For example, if the 3rd input bit of f_c is fixed while the other 4 bits go through all possible values in $\mathbb{F}_2^4$, f_c outputs 10 zeros instead of 8 zeros. To attack, one sends a series of challenges to the reader such that four inputs of f_a (or f_b) are constant while some of the other bits in each challenge vary. The key stream bits are necessarily biased towards either 0 or 1 because of f_c's poor resiliency, which unveils the output bit of f_a (or f_b). Knowing the output bit of f_a (or f_b), the attacker is able to test the inputs to this function. After recovering a number of key bits by repeating this process, exhaustively searching the remaining bits becomes feasible. In fact, Nohl et al. found particularly strong biases in 12 key bits that can be recovered with reasonable effort from the first bit of the key streams. After learning these 12 bits, remaining key bits can be identified within 30 seconds on a single FPGA or within minutes on a typical PC.

Algebraic Attack

A powerful algebraic attack that can recover the entire key of Crypto-1 in 200 seconds on a PC given one known IV and about 50 bits of the key stream is described by Courtois et al. [15]. Algebraic attacks are a new and powerful tool for cryptanalyzing many stream ciphers with linear feedback and nonlinear filtering. The basic idea is to represent Crypto-1 as a system of symbolic Boolean equations that involves the output bits, the key bits, the IV bits, and a large number of intermediate variables, and to solve it with SAT solvers [11, 3, 28].

2.3. Practical Attacks

The algebraic weakness of Crypto-1 undoubtedly exposes this cipher to several attacks. However, what makes breaking MiFare Classic tags even easier and more practical is exploiting the algebraic weakness of Crypto-1 together with the flaws in the authentication protocol and pseudorandom number generator.

Decryption without Secret Key

De Koning Gans, Hoepman and Garcia [23] proposed an attack that exploits the malleability of the cipher to read partial information from MiFare Classic tags, even without knowing details of the encryption algorithm. This attack can be briefly described as follows:

1. Eavesdrop on the communication between a reader and a tag.

2. Impersonate a reader and interact with a genuine tag, expecting that the tag will use the same key stream soon. This happens because the tag uses the same nonce after a short while as mentioned, and the attacker

will completely control the reader's nonce by controlling when to start the communication.

3. By repeating Step 2 many times, the attacker will have $C = P + ks$, and $C' = P' + ks$, and then $P + P'$. If P happens to be known by the attacker, P' can be deduced.

4. By repeating Step 3 many times, the attacker is able to read all blocks of the tag without knowing the secret key.

Fast Recovery of Secret Key

In contrast to the reverse engineering approach, Garcia et al. [22] studied the interaction between tags and readers from a communication perspective, observing in the authentication protocol that (1) since $L^{64}(n_T)$ can be easily computed given n_T, the attacker is able to capture the key stream used in $(L^{64}(n_T) + ks_2)$, and (2) if the tag is manipulated by an attacker and does not respond in Step 3 of the authentication protocol, $(halt + ks_3)$ reveals ks_3 since $halt$ is a constant. These observations shed light on two attacks designed in the same paper.

The first attack relies on the fact that the tag nonce directly manipulates the internal state of the cipher. First, the attacker obtains ks_2 and ks_3 as mentioned. Further, if $n_T = 0x0000???0$, where "0" and "?" represent a nibble of zeros and a nibble of unknown value, the LFSR in Crypto-1 is of the form $0x000?????????$ while feeding in the reader nonce n_R. By looking up $ks_2||ks_3$ in a pre-computed table consisting of tuples $(lfsr, ks)$ where $lfsr$ runs over all values of the form $0x000?????????$ and ks is the corresponding key stream, e.g., $ks_2||ks_3$, the attacker is able to immediately recover the 36-bit internal state of the cipher. In addition, given the state of an LFSR at a certain clock t, one can use the linear relation to compute its previous state at clock $(t - 1)$ in order to recover the secret key. In the second attack, instead of using the table, the attacker can also use the invertibility of the filtering function to recover the state of the LFSR at the end of the authentication. This way, the adversary only needs access to a single authentication session.

Another interesting attack is proposed [10], with a success probability of 0.821 and an online time of 11.7 seconds. The authors make the subtle observation that the three bits of the key stream used for decrypting the last three bits of the encrypted n_R are almost independent of n_R. Consequently, varying n_R while fixing n_T and collecting the corresponding $(n_R + ks_1)$ and $(L^{64}(n_T) + ks_2)$ enable the calculation of the internal state of the LFSR. At

CT-RSA'13, Golić [20] improved this attack by achieving a similar success probability with an online time of only about 3.5 seconds.

Clone a Tag Wirelessly

To clone a tag, the previously-described attacks require that the adversary has access either to an eavesdropped communication session or to a legitimate reader. As such, these attacks cannot be performed undetected from a system integrator's perspective. Four devastating attacks are demonstrated [25] that can be executed with only wireless access to victim tags, and they are fast enough to clone victim tags almost immediately. We summarize these attacks in Table 1 and list the two key ideas upon which attacks are based:

- The tag computes parity bits over the plaintext instead of over the ciphertext, and the key stream bits for encrypting the parity bits are reused to encrypt the next bit of plaintext. Hence, if the attacker challenges a genuine tag with eight random bytes (and eight random parity bits), with a probability of $1/256$, the parity bits are correct and the tag responds with a 4-bit error code, encrypted with the first four bits of ks_3. Taking these 12 leaked bits, the attacker is able to exhaustively search (part of) the secret key.

- When a reader is already communicating with a tag in the second authentication, the tag's challenge is also sent encrypted. Since only 2^{16} values are possible, an attacker can simply guess a nonce to recover 32 bits of key stream. In fact, fewer guesses of the tag's challenge might suffice, as the distance between the tag nonces used in consecutive authentication attempts strongly depends on the time between those attempts.

Table 1. Four Attacks Proposed in [25]

	Data Complexity	Time Complexity	Memory
Brute-force	1500 authentications	0.5 hr	Negligible
Brute-force (varying n_R)	28500 authentications	1 min	Negligible
Brute-force (varying n_T)	4096 authentications	Negligible	384GB
Nested authentication attack [1]	Negligible	1 sec	Negligible

2.4. Discussion

Because worldwide infrastructures with compliance to MiFare Classic cannot be changed instantly, the most effective countermeasure against attacks where a reader is involved is to enable abnormal detection mechanisms on the reader side to make the attacks more expensive, while the countermeasure for the tag-only attacks is to put the cards into electromagnetic-shield covers.

3. Hitag-2 Cipher

As the successor of Crypto-1 cipher, Hitag-2 is an LFSR-based stream cipher that is widely deployed in MiFare Classic tags and commonly used in the automobile industry. This proprietary cryptographic algorithm was kept secret by the manufacturer until it was reverse-engineered in 2007 [39] and 2008 [31]. In this section, we focus on the security of the Hitag-2 cipher, the corresponding authentication protocol, and cryptanalytic techniques developed to break them.

3.1. Specification

Hitag-2 employs a 48-bit LFSR and two layers of nonlinear filtering and produces one bit key stream per clock tick. Besides the 48-bit secret key $k = k_0...k_{47}$ shared between the reader and the tag, Hitag-2 operates on the 32-bit serial number of the tag (which is transmitted in plaintext) and 32-bit nonce supplied by the reader during the authentication.

Initialization

During the initialization, the LFSR loads 32 bits of the serial number id and 16 lower bits of the secret key. During the ith clock cycle, the bit produced by the filtering function is XOR-ed with the $(i+16)$th bit of the key and XOR-ed with the ith bit of the nonce before feeding it back to the LFSR, as shown in Figure 2. Note that the regular feedback bit of the LFSR is not in use during initialization. After 32 clock ticks, the cipher enters the working phase. The nonlinear filtering functions are defined below:

$$
\begin{aligned}
f_a(x_3,x_2,x_1,x_0) &= x_0x_1x_2 + x_0x_2 + x_0x_3 + x_1x_2 + x_0 + x_1 + x_3 + 1 \\
f_b(x_3,x_2,x_1,x_0) &= x_0x_1x_3 + x_0x_2x_3 + x_1x_2x_3 + x_0x_1 + x_0x_2 + x_1x_2 + x_0 \\
&\quad + x_1 + x_3 + 1 \\
f_c(x_4,x_3,x_2,x_1,x_0) &= x_0x_1x_2x_4 + x_0x_1x_3x_4 + x_0x_1 + x_0x_2x_3 + x_0x_3x_4 + x_1x_2x_3 \\
&\quad + x_1x_2x_4 + x_1x_2 + x_1x_3 + x_1x_4 + x_1 + x_2x_3x_4 + x_2x_4 \\
&\quad + x_3x_4 + x_3 + 1.
\end{aligned}
$$

Key Stream Generation

At each clock cycle, the LFSR is updated regularly and the cipher outputs a key stream bit b using the nonlinear filtering function:

$$\begin{aligned} b \;=\; & f_c(f_a(x_1, x_2, x_4, x_5), f_b(x_7, x_{11}, x_{13}, x_{14}), f_b(x_{16}, x_{20}, x_{22}, x_{25}), \\ & f_b(x_{27}, x_{28}, x_{30}, x_{32}), f_a(x_{33}, x_{42}, x_{43}, x_{45})), \end{aligned}$$

where $(x_0...x_{47})$ is the internal state of the LFSR. This process is repeated until the desired number of key stream bits is generated.

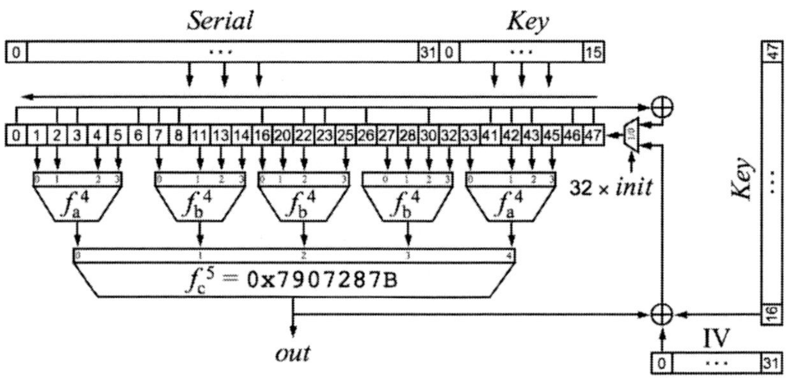

Figure 2. Stream Cipher Hitag-2.

Mutual Authentication Protocol

Similar to Crypto-1, Hitag-2's authentication protocol is tightly coupled with the cipher's design:

1. The reader starts the communication by sending an authenticate command, and the tag answers with its serial number.

2. The reader responds with a 32-bit nonce n_R (encrypted using the first 32 bits of the key stream $ks_0...ks_{31}$) and an authenticator a_R, i.e., the XOR of the next 32 bits of the key stream $ks_{32}...ks_{63}$ with a constant 0x$FFFFFFFF$. Note that $ks_1...ks_{63}$ are produced when Hitag-2 cipher feeds in $n_R + K_{16}...K_{47}$.

3. If the received a_R can be verified, the tag sends back the configuration and password information that is encrypted by the successive key stream bits. From now on, all communication between the reader and the tag is encrypted.

3.2. Cryptanalytic Attacks

Some Observations

One obvious flaw in the design of this cipher is its short key, i.e., 48 bits. While this brute-force process could take up to four years on a Pentium-IV processor, Stembera and Novotny [36] verified that the key can be exhaustively searched in less than two hours on the COPACOBANA2 high-performance cluster of FPGAs. Further, the period of the key stream produced is upper bounded by the period of the LFSR sequence, which is 2^{48}. Therefore, the key stream is repeated after encrypting $32TB$ plaintext and the cipher is vulnerable to time/data/memory tradeoff attacks [7, 38].

Even worse, as the higher 32 bits of the secret key are XOR-ed with the 32 bits of the reader's nonce during initialization, the attacker, by controlling a reader, is able to change the higher 32 bits of the secret key to attack this cipher by merely flipping n_R. For example, several attacks can be mounted to distinguish Hitag-2's ciphertext from a random data stream:

- **Distinguishing with related keys and related nonces**: Let the two related keys differ in the higher 32 bits, i.e., ΔK, and let the two related nonces differ by Δn_R. As long as $\Delta K + \Delta n_R = 0$, the two output streams by the two instances of this cipher should be identical. Apparently, this property is not true for a random oracle and can be used as a distinguisher.

- **Distinguishing with related nonces**: After initialization, the internal states of two running instances of Hitag-2 are strongly correlated if the nonces they consumed differ by one bit. So are the two key streams. This statistical weakness is visualized in Figure 3. As can be seen, if $\Delta n_R = 2^i$, some of the two corresponding output bits are strongly correlated – for example, the first key stream bit from two instances are always equal if $\Delta n_R = 2^{29}$.

Last but not least, a small number of initialization rounds and the lack of nonlinear input during key stream generation decrease overall cryptanalytic strength, since the algebraic degree of each bit of the internal state stays low, which is later exploited in [16, 34].

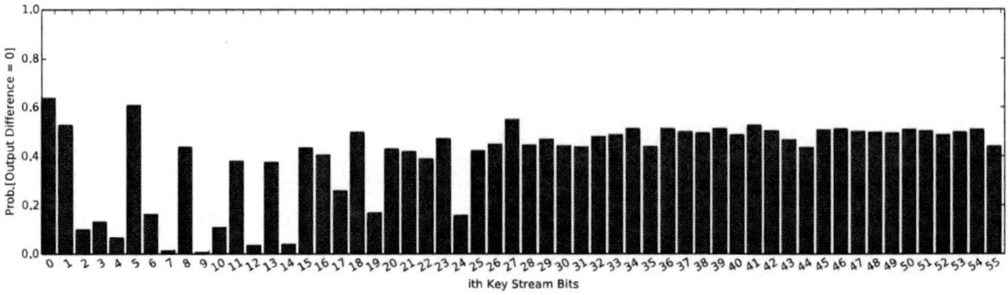

(a) $\Delta n_R = 2^{28}$

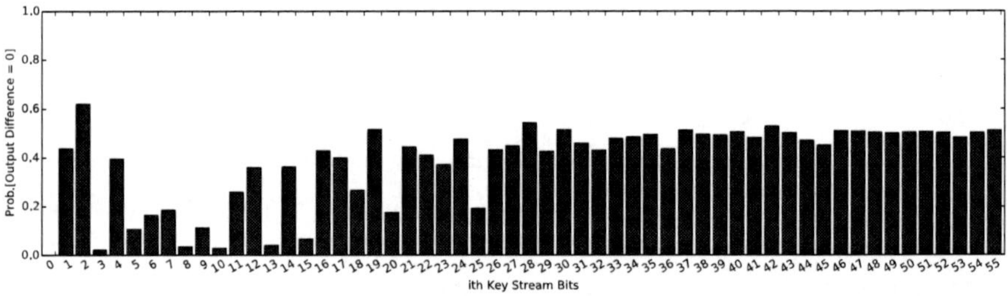

(b) $\Delta n_R = 2^{29}$

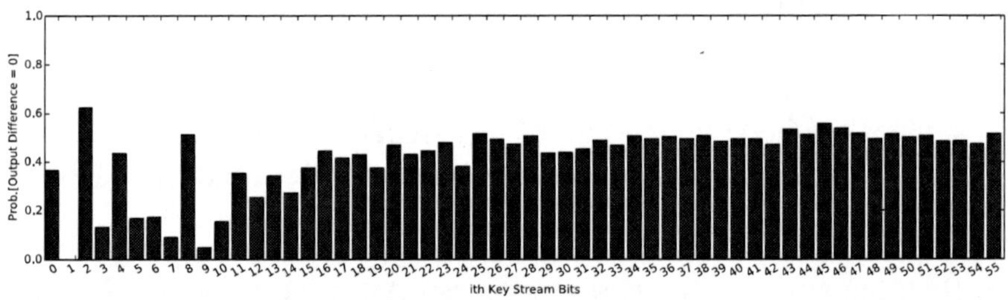

(c) $\Delta n_R = 2^{30}$

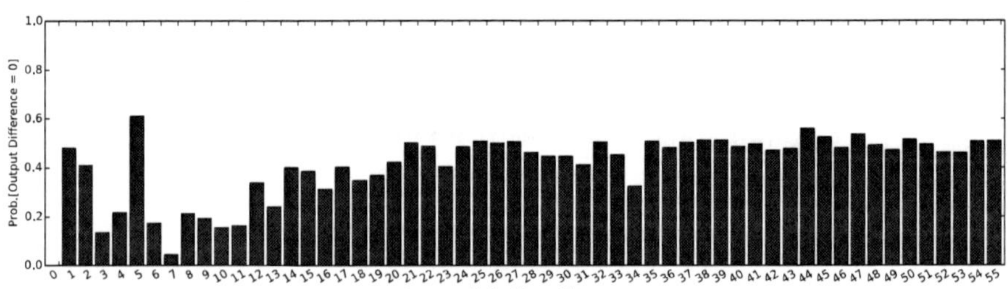

(d) $\Delta n_R = 2^{31}$

Figure 3. Bias in Differentials for First 56 Output Bits If $\Delta n_R = 2^i$.

Algebraic Attacks

The publication of the Hitag-2 cipher received much attention from the security and crypto community. Courtois et al. illustrated an algebraic attack [16] against this cipher. They broke the cipher by writing it as a system of equations with a very large number of unknowns over $\mathbb{F}_2$ and solving the problem with an SAT solver [28]. As verified [16], the algebraic attack takes several hours to recover the key. Unfortunately, this is a black box attack and the researchers could neither find any concrete algebraic weakness of the cipher nor understand why this system of equations is easily solvable. A similar attack was presented in [33] almost at the same time. Shortly after, Soos et al. [37] introduced a series of optimizations on SAT solvers that made it possible to reduce the time to launch algebraic attacks to 40 seconds for Crypto-1 and 6.5 hours for Hitag-2 on a regular PC.

Cube Attack

Cube attack was formally introduced by Ita Dinur and Adi Shamir in Eurocrypt'09 [18] and its principle can be described by a toy example. Consider the following dense Boolean polynomial of degree 3 over three secret variables x_1 x_2, x_3 and three public variables v_1, v_2, v_3:

$$\begin{aligned} P(v_1, v_2, v_3, x_1, x_2, x_3) = v_1v_2v_3 + v_1v_2x_1 + v_1v_3x_1 + v_2v_3x_1 \\ +v_1v_2x_3 + v_1v_3x_2 + v_2v_3x_2 + \\ v_1v_3x_3 + v_1x_1x_3 + v_3x_2x_3 + x_1x_2x_3 + v_1v_2 + v_1x_3 + v_3x_1 + x_1x_2 \\ +x_2x_3 + x_2 + v_1 + v_3 + 1 \end{aligned}$$

By summing up $P(v_1, 0, 0, x_1, x_2, x_3)$, $P(v_1, 0, 1, x_1, x_2, x_3)$, $P(v_1, 1, 0, x_1, x_2, x_3)$, $P(v_1, 1, 1, x_1, x_2, x_3)$, we get

$$\begin{aligned} P_{s(I)} &= P(v_1, 0, 0, x_1, x_2, x_3) + P(v_1, 0, 1, x_1, x_2, x_3) \\ &\quad +P(v_1, 1, 0, x_1, x_2, x_3) + P(v_1, 1, 1, x_1, x_2, x_3) \\ &= x_1 + x_2 + v_1, \end{aligned}$$

where $I = \{1, 2\}$ is the index set. Since $P_{s(I)}$ and v_1 are known, the attacker is able to get one linear relation regarding two secret variables. Similarly we can obtain $P_{s(I)} = x_1 + x_2 + x_3 + v_2$ when $I = \{1, 3\}$ and $P_{s(I)} = x_1 + x_3 + v_3$ when $I = \{1, 2\}$. Consequently, the master polynomial can be solved as a simple system of three linear equations in the three secret variables by a chosen message attack that evaluates it for just eight combinations of values of its public variables.

In [34], the authors identified the cubes of 32 bits of the secret key separately, a few of which are listed in Table 2. The remaining key bits can be exhaustively searched. The experiments show that the 48-bit secret key of Hitag-2 can be recovered within one minute on a PC.

Table 2. Some Cubes Found in [34]

Secret Variables	Index Set	Key Stream Bit	Secret Variables	Index Set	Key Stream Bit
K_6	4, 15	0	K_{12}	10, 12, 16	0
K_3	7, 10	0	$K_0 + K_{13}$	5, 6, 23	1
K_0	5, 7	1	K_{27}	4, 10, 24	2
K_5	3, 13	2	K_{25}	1, 13, 8	3
K_8	3, 14	2	K_{14}	2, 7, 31	3
K_{17}	4, 6	2	$K_1 + K_{10}$	3, 31, 13	4
K_{21}	1, 15, 16	0	K_7	6, 8, 31	4
$K_2 + K_8$	4, 15, 19	0	K_{22}	3, 6, 8	6
...	...	...	...	...	...

3.3. Practical Attacks

Although many cryptanalytic vulnerabilities have been identified, they can hardly be used in a real-world scenario because the chosen-IV capability, a common assumption of all aforementioned cryptanalytic attacks, is prohibited by Hitag-2's authentication protocol, unintentionally. This is because of another design flaw such that the tag contributes no randomness to the authenticator. As a consequence, this protocol allows an attacker to replay any valid n_R and a_R to victim tags to pass the authentication.

A Real Encryption Oracle

Verdult et al. [38] presented an impressive technique that transforms a genuine tag to an encryption oracle, and the attacker is able to use this encryption oracle to produce a key stream of arbitrary length without knowing the secret key. This attack works as follows:

1. The attacker passes the tag's authentication by replaying a valid n_R and a_R.
2. The attacker guesses 10 bits, i.e., $cmd(11, 0, 0)$, which is the command to read block 0 on the tag with 0 redundant message, which is encrypted.

If the tag's response is 32-bits long, i.e., id XOR-ed with key stream bits, the attacker can be sure that the guessed encrypted command is correct. Since $cmd(11, 0, 0)$ and id are known, the attacker obtains 40 bits of the ciphertext.

3. The attacker restarts the communication, replays to pass the authentication and uses the obtained 40 bits to encrypt a 40-bit command, i.e., $cmd(11, 6, 0)$, which means to read block 0 on the tag with six redundant messages. Since this command is encrypted correctly, the attacker obtains another 32 bits of ciphertext used to encrypt id from the tag's response.

4. By repeating Step 3 multiple times and increasing the length of the commands by appending even more redundant messages, the attacker could capture a key stream of arbitrary length.

Fast Recovery of Secret Key

Moreover, Verdult et al. [38] demonstrated a key recovery attack that takes only 5 minutes and requires only 136 (n_R, a_R) pairs. This attack exploits the following two properties in Hitag-2's protocol stack and efficiently guesses and determines the first 34 bits of the key before brute forcing the rest. We ignore the detailed steps here.

- With probability 0.25, the output of the filtering function f is determined by the 34 leftmost bits of the internal state.

- At the end of the initialization, the first 16 bits of the internal state (which are actually $k_0, ...k_{15}$) do not depend on n_R.

3.4. Discussion

Many serious vulnerabilities in the Hitag-2 and its usage in the automotive industry have been discovered. Among them, replaying to pass the authentication (and stealing cars protected by this cipher) might be the most devastating attack – it is extremely easy to launch but not trivial at all to eliminate. All the evidence suggests that this cipher and its protocol should be deprecated.

4. The Atmel Cipher

A more recent proprietary stream cipher, which is called the Atmel Cipher by crypto researchers [26, 5], has been employed in the Atmel chip families SecureMemory (introduced in 1999), CryptoMemory (introduced in 2002), and CryptoRF (introduced in 2003) to guarantee authenticity, confidentiality, and integrity. While the CryptoRF family and CryptoMemory family use exactly the same cipher, the SecureMemory family uses a simplified version of the cipher in the sense that it does not memorize the last eight bits of the output and does not feed them back into the shift registers. These families of chips are used mainly in access cards, health care cards, loyalty cards, internet kiosk, energy meters, and e-government, as well as print cartridges, NVIDIA graphics cards, removable storage devices, counterfeit protection and so on. In this section, we focus on the security of the Atmel Cipher and the corresponding authentication protocol and survey some cryptanalytic techniques developed to break this cipher.

4.1. Specification

The Atmel Cipher has an internal state of 117 bits (109 bits for SecureMemory), organized in four shift registers (three for SecureMemory). Every clock cycle, two of the registers produce 4 bits respectively, while the output of the third is used to select bits from one of the two other registers to produce 4 bits of the key stream. Let's denote the left register as $l = (l_0 \ldots l_6) \in \mathbb{F}_{2^5}^7$, the middle register as $m = (m_0 \ldots m_6) \in \mathbb{F}_{2^7}^7$, the right register as $r = (r_0 \ldots r_4) \in \mathbb{F}_{2^5}^5$ and the feedback register as $f = (f_0 f_1) \in \mathbb{F}_{2^4}^2$. At each clock cycle, one input $a \in \mathbb{F}_{2^8}$ is injected to the three registers as follows:

$$l_2 \leftarrow l_2 + b_3 b_4 b_5 b_6 b_7 \qquad m_4 \leftarrow m_4 + b_4 b_5 b_6 b_7 b_0 b_1 b_2$$
$$r_1 \leftarrow r_1 + b_0 b_1 b_2 b_3 b_4$$

where $b = (b_0 \ldots b_7) = a + f_0 f_1$ for CryptoMemory/CryptoRF and $b = a$ for SecureMemory. Then the three registers are converted into successor states by shifting right and producing the new states as below:

$$l_0 \leftarrow l_3 \boxplus L(l_6) \qquad m_0 \leftarrow m_5 \boxplus L(m_6) \qquad r_0 \leftarrow r_2 \boxplus L(r_4)$$

where L denotes the 1-bit left rotation and the modified modular addition is defined as follows:

$$x \boxplus y = \begin{cases} x + y \bmod 2^n - 1 & \text{if } x = y = 0 \text{ or } x + y \neq 0 \\ 2^n - 1 & \text{otherwise} \end{cases}$$

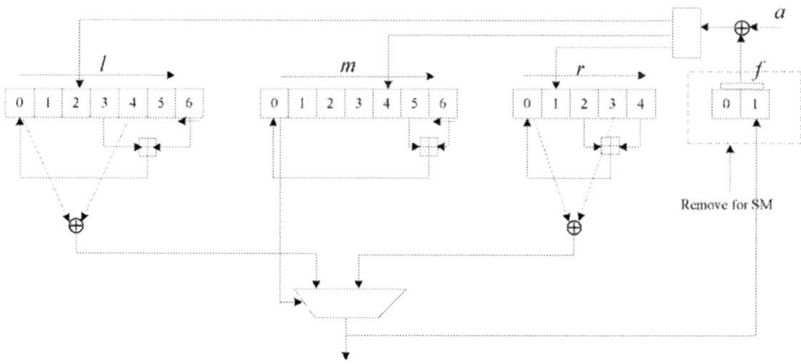

Figure 4. Stream Cipher Atmel.

where $+$ is the integer addition.

The 4-bit output is determined by the rightmost four bits of m_0 in a bit-by-bit manner. For $i \in \{0, 3\}$, if $m_{0,i+3} = 0$, the ith bit of $(l_0 + l_4)$ is selected; otherwise the ith bit of $(r_0 + r_3)$ is selected. For CryptoMemory/CryptoRF, $f_0 = f_1$ and f_1 is set to the output nibble at the end of this clock cycle.

As a notational aid, we define the transition function suc that takes an input a and the current state s and outputs the successor state $s' = suc(a, s)$, and $suc^n(a, s) = suc^{n-1}(a, suc(a, s))$.

Mutual Authentication Protocol

The authentication protocol is quite standard: the tag sends a 64-bit nonce n_T to the reader and the reader responds with another 64-bit nonce n_R and an authenticator a_R. If a_R can be verified, the tag responds with another authenticator a_T. The reader authenticates the tag by verifying a_T.

Note that a_T and a_R are parts of the key stream produced after feeding in n_T and n_R. In SecureMemory, every second output nibble is discarded and the key stream bits used as a_T and a_R are interleaved. Table 3 shows schematically how to assemble the authenticators using the nibbles from the key stream, where "–" indicates the current nibble is unused.

Initialization

The cipher is initialized during the authentication protocol. At the beginning of the protocol, the internal state has all components equal to zero and evolves from there. After feeding in the inputs according to Table 4, the cipher finishes its initialization.

Table 3. Authenticators

SecureMemory																					
–	$a_{T,0}$	–	$a_{T,1}$	–	$a_{R,0}$	–	$a_{R,1}$	–	$a_{T,2}$	–	$a_{T,3}$	–	$a_{R,2}$	–	$a_{R,3}$	–	$a_{T,4}$	–	$a_{T,5}$	–	$a_{R,4}$
–	$a_{R,5}$	–	$a_{T,6}$	–	$a_{T,7}$	–	$a_{R,6}$	–	$a_{R,7}$	–	$a_{T,8}$	–	$a_{T,9}$	–	$a_{R,8}$	–	$a_{R,9}$	–	$a_{T,10}$	–	$a_{T,11}$
–	$a_{R,10}$	–	$a_{R,11}$	–	$a_{T,12}$	–	$a_{T,13}$	–	$a_{R,12}$	–	$a_{R,13}$	–	$a_{T,14}$	–	$a_{T,15}$	–	$a_{R,14}$	–	$a_{R,15}$		
CryptoMemory/CryptoRF																					
–	–	–	–	$a_{R,0}$	$a_{R,1}$	–	–	–	–	–	$a_{R,2}$	$a_{R,3}$	–	–	–	–	–	$a_{R,4}$	$a_{R,5}$	–	–
–	–	–	$a_{R,6}$	$a_{R,7}$	–	–	–	–	–	$a_{R,8}$	$a_{R,9}$	–	–	–	–	–	$a_{R,10}$	$a_{R,11}$	–	–	–
–	–	$a_{R,12}$	$a_{R,13}$	–	–	–	–	–	$a_{R,14}$	$a_{R,15}$	$a_{T,2}$	$a_{T,3}$	$a_{T,4}$	$a_{T,5}$	$a_{T,6}$	$a_{T,7}$	$a_{T,8}$	$a_{T,9}$	$a_{T,10}$	$a_{T,11}$	$a_{T,12}$
$a_{T,13}$	$a_{T,14}$	$a_{T,15}$																			

Table 4. Variables Feeding In During Initialization

SecureMemory																							
n_{T_0}	n_{T_1}	n_{R_0}	n_{T_2}	n_{T_3}	n_{R_1}	n_{T_4}	n_{T_5}	n_{R_2}	n_{T_6}	n_{T_7}	n_{R_3}	k_0	k_1	n_{R_4}	k_2	k_3	n_{R_5}	k_4	k_5	n_{R_6}	k_6	k_7	n_{R_7}
CryptoMemory/CryptoRF																							
n_{T_0}	n_{T_0}	n_{T_0}	n_{T_1}	n_{T_1}	n_{T_1}	n_{R_0}	n_{T_2}	n_{T_2}	n_{T_2}	n_{T_2}	n_{T_3}	n_{T_3}	n_{T_3}	n_{R_1}	n_{T_4}	n_{T_4}	n_{T_4}	n_{T_5}	n_{T_5}	n_{T_5}	n_{R_2}	n_{T_6}	n_{T_6}
n_{T_6}	n_{T_7}	n_{T_7}	n_{T_7}	n_{R_3}	k_0	k_0	k_0	k_1	k_1	k_1	n_{R_4}	k_2	k_2	k_2	k_3	k_3	k_3	n_{R_5}	k_4	k_4	k_4	k_5	k_5
k_5	k_5	n_{R_6}	k_6	k_6	k_6	k_7	k_7	n_{R_7}															

4.2. Cryptanalytic Attacks

In 2010, Garcia et al. [26] reverse engineered the security mechanisms of the SecureMemory and CryptoMemory/CryptoRF, including the ciphers, the authentication protocols, and the password mechanisms. In their paper, two algorithms to recover the secret key are given: the first algorithm recovers the secret key of SecureMemory with probability 0.57 and with time complexity $2^{39.4}$ and 2640 plaintext/ciphertext pairs; the second algorithm recovers the secret key of CryptoMemory/CryptoRF with time complexity of 2^{58}. Since they are similar, we concentrate on the first one.

There are two weaknesses in the SecureMemory cipher that make the attacks possible: (1) there is a high correlation between the output of the right (or left) register and the key stream, i.e., with probability 0.75, an output bit of the right (or left) register equals the corresponding key stream bit; (2) the three registers operate independently, and can be tackled in a divide-and-conquer manner. As such, the attack designed to recover the secret key of SecureMemory works as follows:

1. Eavesdrop one authentication session to obtain n_R, n_T, a_R and a_T.

2. For each of all the possible 2^{25} initial states of the right register, feed in n_R and n_T, and run r independently of l and m to generate 128 bits as specified. If the guessed initial state is correct, 96 of the 128 bits should match the corresponding bits in a_R and a_T; otherwise, only 64 bits will match. Approximately, with probability 0.57, the correct guess matches more bits than all the wrong ones.

3. After recovering the partial key used in the right register, the attacker can conclude that the "unmatched" 32 bits in the key stream are contributed by the left register. By enumerating all 2^{35} possible initial states of the left register and discarding the ones that cannot produce the "unmatched" 32 bits, the attacker is able to narrow down the number of candidate keys to several hundred.

4 At last, a meet-in-the-middle attack is launched to recover the internal state of the middle register.

In 2011, Biryukov, Kizhvatov and Zhang discovered faster and more data-efficient attacks [5] that recover the secret key of SecureMemory with probability 0.75 with time complexity $2^{29.8}$ and recover the secret key of CryptoMemory/CryptoRF with time complexity 2^{50}. Their attacks work similar to the previous one in Step 1 and Step 2. However, Biryukov et al. noticed that the observed nibbles from l only depend on part of l's initial state. Therefore, guessing and determining the partial key using a Viterbi-decoding-like algorithm in l is sufficient and speeds up entire attacks.

4.3. A Practical Attack via Differential Power Analysis

Previous work has already pointed out cryptographic weaknesses in the Atmel Cipher, but the most efficient attack still requires substantial computational effort. Balasch et al. [12] focused on power analysis attacks as they are usually more practical if the attacker either possesses the victim device or has physical access to it. Surprisingly, Balasch et al. found that the Atmel Cipher (at least for CryptoMemory) is not protected by countermeasures against differential power analysis (DPA) and the entire key can be extracted given a sufficient number of power measurements.

DPA is based on the relationship between the power consumption and the Hamming weight of the data manipulated at a given point in time. To perform DPA, one initially runs the encryption on the targeted hardware platform a couple of times with a fixed key and distinct plaintexts, which produces the different ciphertexts and results in different power consumption. In their experiments, they obtained a set of 1000 power traces sampled during executions of the mutual authentication protocol with random n_Rs. They extracted the peaks highlighted in Figure 5 (a). Unlike typical attacks on block ciphers where parts of the key cannot be recovered independently, here parts of the key can be recovered in the order in which they are fed into the cipher. Finally the correlation between the partial key, e.g., $k_6 \in \mathbb{F}_{2^8}$, and the power consumption can be identified as in Figure 5 (b), where the correlation coefficients, computed over all traces collected for all 256 hypotheses, are plotted and, apparently, the spike indicates the correct k_6. Similarly, k_7, k_8, ..., can be recovered subsequently.

In all, it is surprising that Atmel's crypto chips lack even the countermeasures against basic, well-understood power analysis attacks. Additionally, algebraic side-channel analysis such as [32] would perhaps allow these attacks to work with even fewer traces.

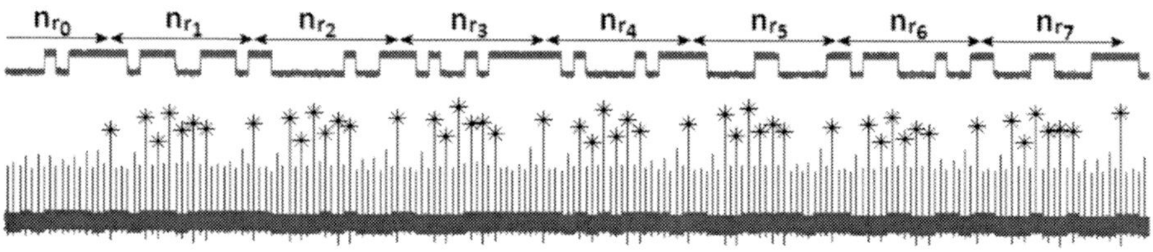

(a) Power Traces of the cipher when feeding in n_R. Interesting peaks are marked with *.

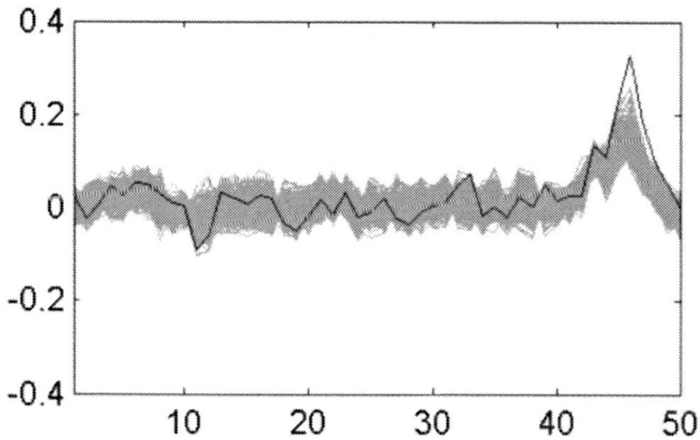

(b) Correlation coefficients per hypothesis for k_6

Figure 5. Differential Power Analysis of Atmel Cipher in [12].

4.4. Discussion

These attacks once again suggest that such a proprietary cipher is usually flawed. Fortunately, although the simplified version of the Atmel Cipher is completely broken from a practical point of view, the more full-fledged version still provides a certain degree of security. Future research exploiting design details of these chips is likely to find more practical attacks according to the history of attacks on Crypto-1 and Hitag-2.

5. A2U2: A Lightweight Stream Cipher for RFID Tags

More recently, David et al. [17] proposed another stream cipher called A2U2, which is the most lightweight cipher proposed so far, in terms of its hardware footprint of 284 GE (Gate Equivalent). Besides its compactness, the throughput of A2U2 is approximately five times greater than that of PRESENT, KTANTAN32 and PRINT. The security analysis of A2U2 in the same paper shows: (1) The output sequence of A2U2 can pass the NIST's statistical tests for pseudorandom number generators; (2) The period and the linear complexity of the output sequence are around 2^{70}; (3) Particular attacks are thwarted since variable number of clock cycles used for the initialization ensures that the cipher outputs different ciphertexts for identical plaintexts.

In this section, we address the security of the lightweight stream cipher A2U2. Our cryptanalytic results show that A2U2 is insecure under a simple chosen-plaintext attack, which enables the full key recovery of A2U2 by executing two encryptions with particular plaintexts of 653 bits and solving 32 sparse systems of linear equations (where each system has 56 unknowns and around 28 unknowns can be directly obtained without computation) in about 0.16 second on a standard PC.

5.1. Specification

As illustrated in Figure 6, the stream cipher A2U2 is composed of four building blocks: a 7-stage LFSR, a combination of two nonlinear feedback shift registers (NFLRs), a key schedule module, and a filtering function.

After a 61-bit secret key $(k_0, ..., k_{60})$ is burnt into an RFID tag, the tag is capable of encrypting plaintext bits $p_\delta, \ldots, p_n$ to the corresponding ciphertext bits $c_\delta, \ldots, c_n$, where δ is the number of clock cycles required to initialize the internal state of A2U2 and is determined by the following procedure:

1. The RFID reader and the tag generate and exchange two 32-bit random numbers $RND_R = (a_0, \ldots, a_{31}) \in \mathbb{F}_2^{32}$ and $RND_T = (b_0, \ldots, b_{31}) \in \mathbb{F}_2^{32}$, respectively;

2. The value $(RND_R + RND_T)$ is then loaded into the LFSR and two NFSRs of the A2U2 cipher;

3. Both the LFSR and two NFSRs run δ clock cycles for initialization without any output until the state of the LFSR reaches all ones. From the next clock cycle, the stream cipher A2U2 outputs the ciphertext bits.

The details of the building blocks of A2U2 are given below.

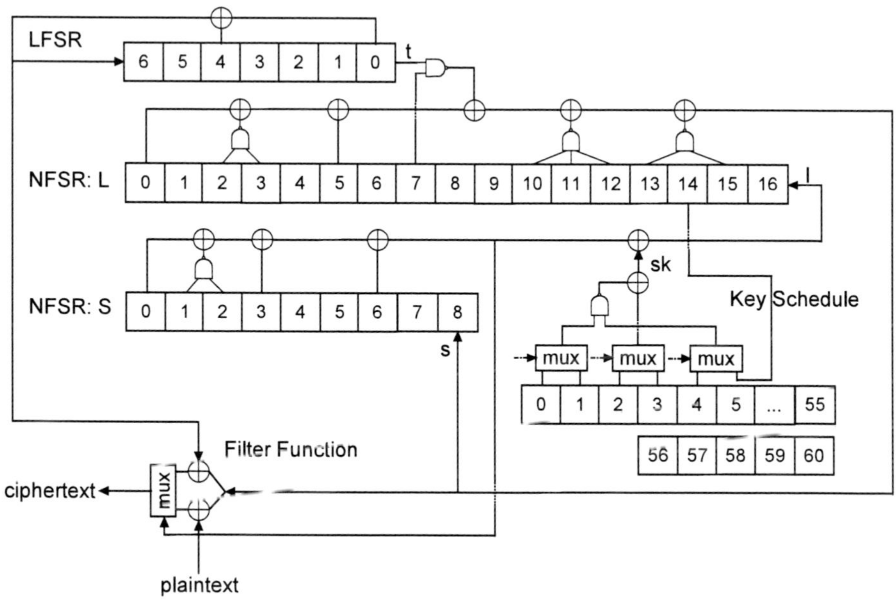

Figure 6. Stream Cipher A2U2.

LFSR

The LFSR has seven stages denoted by a binary vector $(t_{i+6}, \ldots, t_i) \in \mathbb{F}_2^7, i = 0, \ldots, n$. Moreover, the following recursive relation holds:

$$t_{i+7} = t_i + t_{i+4}, \text{ for } i = 0, \ldots, n.$$

Note that the generated sequence is an m-sequence [21] with a maximum period $2^7 - 1 = 127$. In the above Step 2, $((a_0, \ldots, a_4) + (b_0, \ldots, b_4) + (k_{56}, \ldots, k_{60}))$ is loaded into $(t_4, \ldots, t_0)$ of the LFSR, while leaving t_5 to be a constant 1 and t_6 to be a constant 0.

Two NFSRs

This block borrows part of the design from the lightweight block cipher KATAN/KTANTAN [13]. For $i = 0, \ldots, n$, the feedback functions can be represented as follows:

$$\begin{aligned} s_{i+8} &= l_i + l_{i+2} \odot l_{i+3} + l_{i+5} + l_{i+7} \odot t_{i+6} \\ &\quad + l_{i+10} \odot l_{i+11} \odot l_{i+12} + l_{i+13} \odot l_{i+15}, \\ l_{i+16} &= s_i + s_{i+1} \odot s_{i+2} + s_{i+3} + s_{i+6} + sk_i, \end{aligned} \tag{1}$$

where sk_i is the subkey bit generated by the key schedule.

Key Schedule

This module derives a sequence of subkeys $(sk_0, \ldots, sk_n)$ from a portion of the secret key $(k_0, ..., k_{55})$, where sk_i is used by the NFSR during the ith clock cycle and is computed as follows:

$$\begin{aligned} sk_i &= mux(k_{mod(5i,56)}, k_{mod(5i+1,56)}, t_{i+1}) \odot \\ &\quad mux(k_{mod(5i+4,56)}, l_{i+14}, t_{i+5}) + \\ &\quad mux(k_{mod(5i+2,56)}, k_{mod(5i+3,56)}, t_{i+3}), \end{aligned} \tag{2}$$

where $mod(x, y)$ returns x modulo y provided that x, y are non-negative integers and $mux()$ is a multiplexer such that, given $x, y, z \in \{0, 1\}$, $mux(x, y, z) = x$ iff $z = 0$ or $mux(x, y, z) = y$ iff $z = 1$.

Filtering Function

The filtering function is essentially a variant of the stop-and-go sequence generator [6], which replaces the XOR operation of key stream bits and plaintext bits in a classical stream cipher, i.e.,

$$c_i = \begin{cases} s_{i+8} + t_i, & \text{if } l_{i+16} = 0, \\ s_{i+8} + p_{f(i)}, & \text{if } l_{i+16} = 1, \end{cases} \quad \text{for } i = \delta, \ldots, \hat{n} \tag{3}$$

where p_i is the plaintext bit fed into the A2U2 cipher during ith clock cycle, c_i is the corresponding ciphertext bit, $f(i) = \delta + \sum_{j=\delta+16}^{i+16} l_j$, $\delta \le i \le n$, $f(\delta) = \delta$, and $\hat{n}$ equals the sum of n and the number of inserted bits from the m-sequence.

5.2. A Lightweight Key Recover Attack

Attacker Model

We consider the classical chosen-plaintext attack against the stream cipher A2U2 implemented on an RFID tag with a fixed high-entropy 61-bit secret key burnt inside. An attacker, equipped with a programmable RFID reader, queries the victim tag with a particular random number RND_R and plaintext bits $p_\delta, ..., p_n$. The attacker's goal is to recover the secret key $(k_0, ..., k_{60})$. Note that in our attack, the reader, manipulated by the attacker, can always adaptively choose RND_R to make $(RND_R + RND_T)$ a constant, e.g., $RND_R + RND_T = 0$.

Step 1. Recover Sequences of s_{i+8} And l_{i+16}

A cryptographic primitive is only as strong as its weakest module, which is the general principle to break a cryptosystem. We noticed that the filter function in A2U2 is quite weak, which enables an attacker to easily recover the internal state of the NFSRs using the following steps.

1. At the δth clock cycle, the LFSR reaches the all-one state and A2U2 starts the ciphertext output. As a result, the attacker knows the sequence $(t_\delta, t_{\delta+1}, ..., t_n)$, which is just a repetition of the m-sequence with period 127 as shown in Table 5.

Table 5. One Period of $(t_\delta, t_{\delta+1}, ..., t_n)$

1, 1, 1, 1, 1, 1, 1, 0, 0, 0, 1, 1, 1, 0, 1, 1, 0, 0, 0, 1, 0, 1, 0,
0, 1, 0, 1, 1, 1, 1, 1, 0, 1, 0, 1, 0, 1, 0, 0, 0, 0, 1, 0, 1, 1, 0,
1, 1, 1, 1, 0, 0, 1, 1, 1, 0, 0, 1, 0, 1, 0, 1, 1, 0, 0, 1, 1, 0, 0,
0, 0, 0, 1, 1, 0, 1, 1, 0, 1, 0, 1, 1, 1, 0, 1, 0, 0, 0, 1, 1, 0, 0,
1, 0, 0, 0, 1, 0, 0, 0, 0, 0, 0, 1, 0, 0, 1, 0, 0, 1, 1, 0, 1, 0, 0,
1, 1, 1, 1, 0, 1, 1, 1, 0, 0, 0, 0.

2. The attacker chooses two complimentary plaintexts $(p_\delta, \ldots, p_n)$ and $(p'_\delta, \ldots, p'_n)$ for encryption, i.e., $p_i + p'_i = 1, \delta \leq i \leq n$. Letting the corresponding ciphertexts be $(c_\delta, \ldots, c_{\hat{n}})$ and $(c'_\delta, \ldots, c'_{\hat{n}})$, we have

$$c_i + c'_i = \begin{cases} 0, & \text{if } l_{i+16} = 0, \\ 1, & \text{if } l_{i+16} = 1, \end{cases} \quad \text{for } i = \delta, \ldots, \hat{n},$$

which reveals the sequence $\{l_{\delta+16}, \ldots, l_{\hat{n}+16}\}$. Consequently, the sequence $\{s_{\delta+8}, \ldots, s_{\hat{n}+8}\}$ can be simply recovered from Eq. (3) once each l_{i+16} is known.

Step 2. Recover Internal States of NFSRs and Subkey sk_i

Given s_{i+8} and l_{i+16} ($i \geq \delta$) obtained in Step 1, the internal states of the two NFSRs are completely exposed to the attacker after $\max(\delta+9, \delta+17) = \delta+17$ clock cycles. Next, the attacker employs the following relation derived from Eq. (1) to recover the subkey bits sk_i for $i = \delta + 17, \ldots, n$:

$$sk_i = s_i + s_{i+1} \odot s_{i+2} + s_{i+3} + s_{i+6} + l_{i+16}.$$

We would like to point out that the attacker is already capable of decrypting any ciphertext with the obtained subkeys $(sk_{\delta+17}, \ldots, sk_n)$. However, the attacker can do even better by fully recovering the secret key as described in Step 3.

Step 3. Fully Recover Secret Key

Without loss of generality, we fix δ to be a specific integer by guessing, e.g., assume $\delta = 88$ hereafter, and first recover the partial secret key bits $k_0, ..., k_{55}$. To this end, the attacker generates $(n - \delta - 16)$ equations (see below) from Eq. (2) as well as the internal state l_i of the NFSRs, and derives the subkeys sk_i, when $i \geq \delta + 17$.

$$\begin{array}{rr}
k_{22} \cdot l_{119} + k_{23} = sk_{105} + 1 & k_{26} \cdot k_{30} + k_{29} = sk_{106} + 1 \\
k_{31} \cdot l_{121} + k_{34} = sk_{107} + 1 & k_{37} \cdot k_{40} + k_{38} = sk_{108} + 1 \\
k_{42} \cdot k_{45} + k_{44} = sk_{109} + 1 & k_{46} \cdot l_{124} + k_{48} = sk_{110} + 1 \\
k_{52} \cdot l_{125} + k_{53} = sk_{111} + 1 & k_0 \cdot l_{126} + k_3 = sk_{112} + 1 \\
k_5 \cdot l_{127} + k_8 = sk_{113} + 1 & k_{11} \cdot k_{14} + k_{13} = sk_{114} + 1 \\
\cdots & \cdots
\end{array}$$

Among these equations, approximately half of them are of the type

$$k_x \cdot l_y + k_z = sk_w + 1, \tag{4}$$

where x, y, z and w are integers. Since l_y is known to the attacker, he can select 56 such equations to form a sparse linear system with full rank and solve it to get $(k_0, ..., k_{55})$. Through extensive experiments, we found that when $n = 512 + \delta$, the attacker can obtain 56 linear independent equations out of 249 Eq. (4)-like equations with probability 1, which implies that in a practical attack scenario the attacker should query the RFID tag with plaintexts of $n = 512 + \delta = 638$ bits (recall that $\delta \leq 126$).

To recover the rest of the key bits $k_{56}, \ldots, k_{60}$, we make use of the fact that δ is indeed the number of clock cycles required to transit the LFSR's state from $(k_{60}, \ldots, k_{56}, 1, 0)$ to $(1, 1, 1, 1, 1, 1, 1)$ or the reverse direction (here we assume $RND_R + RND_T = 0$ for simplicity). In our example, transiting the state $(1, 1, 1, 1, 1, 1, 1)$ reversely for $\delta = 88$ clock cycles gives us $(k_{56}, ..., k_{60}) = (1, 0, 1, 0, 0)$. Due to the uncertainty of δ, the attacker could have 32 possible keys such that the recovered $(k_0, ..., k_{55})$ is a δ-bit shifted version of the right key and $(k_{56}, ..., k_{60})$ is the state determined by δ as listed in Table 6. The attacker then utilizes the obtained one plaintext/ciphertext pair to test all 32 key candidates locally for retrieving the correct one.

Table 6. Deterministic Relation Between δ and $(k_{56}, \ldots, k_{60})$

$(k_{56}, \ldots, k_{60})$	δ	$(k_{56}, \ldots, k_{60})$	δ	$(k_{56}, \ldots, k_{60})$	δ
$(0,0,0,0,0)$	29	$(0,0,0,0,1)$	91	$(0,0,0,1,0)$	36
$(0,0,0,1,1)$	113	$(0,0,1,0,0)$	26	$(0,0,1,0,1)$	108
$(0,0,1,1,0)$	40	$(0,0,1,1,1)$	75	$(0,1,0,0,0)$	111
$(0,1,0,0,1)$	73	$(0,1,0,1,0)$	96	$(0,1,0,1,1)$	98
$(0,1,1,0,0)$	20	$(0,1,1,0,1)$	54	$(0,1,1,1,0)$	48
$(0,1,1,1,1)$	100	$(1,0,0,0,0)$	59	$(1,0,0,0,1)$	43
$(1,0,0,1,0)$	22	$(1,0,0,1,1)$	66	$(1,0,1,0,0)$	88
$(1,0,1,0,1)$	70	$(1,0,1,1,0)$	56	$(1,0,1,1,1)$	117
$(1,1,0,0,0)$	120	$(1,1,0,0,1)$	78	$(1,1,0,1,0)$	50
$(1,1,0,1,1)$	10	$(1,1,1,0,0)$	14	$(1,1,1,0,1)$	83
$(1,1,1,1,0)$	102	$(1,1,1,1,1)$	126		

Computational Complexity of the Attack

Our attack is an ultra-efficient chosen-plaintext attack in terms of the computational overhead. Table 7 summarizes the computational complexity of the proposed attack.

Table 7. Computational Complexity of the Proposed Attack

Recovery Bits		Computation Cost
s_i	$i = \delta + 0, \ldots, n$	one encryption of n bits
l_i	$i = \delta + 17, \ldots, n$	one encryption of n bits
sk_i	$i = \delta + 17, \ldots, n$	negligible
k_i	$i = 0, \ldots, 60$	solve 32 sparse systems of linear equations $\approx$ 1 second on a standard PC

Generally speaking, solving a system of linear equations with m variables requires $O(m^3)$ steps by the Gaussian elimination. Nonetheless, as observed in our experiment, the linear system in question is quite special such that approximately 28 equations are of type

$$k_z = sk_w + 1,$$

which immediately returns the key bits. Moreover, the remaining equations can be solved in around 0.005 second on a standard PC in our testing. In the worst case, the attacker has to solve 32 such systems of linear equations using around 0.16 second, which is negligible for the attacker.

5.3. Summary

We identify in this section the security vulnerabilities of the A2U2 lightweight stream cipher and develop an ultra-efficient chosen-plaintext attack to fully recover the secret key of A2U2 through querying the encryption function twice on the victim tag and solving 32 sparse systems of linear equations in around 0.16 second. Our cryptanalysis implies that A2U2 has been completely broken and is not eligible to provide confidentiality and authenticity for RFID communication, which settled the concerns that are made in the conclusion of the original A2U2 proposal [17].

6. Conclusion

In this chapter, we present many cryptanalytic techniques and practical attacks developed for the most common lightweight stream ciphers employed in the industry. These ciphers, together with their corresponding authentication protocols, are meaningfully compromised and should be retired.

Nevertheless, there are four important lessons to be learned from the rise and fall of these ciphers. First of all, lightweight ciphers should not be designed to provide lightweight security, e.g., using 48-bit keys. Security can only be true or false, and a somewhat-secure design is very likely to become a flawed design as time goes by. Second, lightweight ciphers should not be designed merely by removing security margins due to the resource constraints. For example, lack of sufficient rounds of initialization or rushing to output the key stream contributes to the vulnerability of these ciphers. Instead, finding and employing more efficient functions, structures and mechanisms for lightweight crypto primitives should be encouraged equally. The third lesson is that all ciphers suffer from aging, especially lightweight ciphers designed with little or no security margin. There are two reasons for cipher aging. Most prominent is the development of cryptanalytic techniques. For example, the first three ciphers discussed in this chapter were designed at a time when differential cryptanalysis and linear cryptanalysis received a lot of attention. Consequently, these ciphers, while providing resistance to differential/linear cryptanalysis to some degree, are extremely vulnerable to cryptanalytic techniques that appeared later – such as the algebraic attack and cube attack. Another reason is underestimating the advancement of computation power – attacks like exhaustive search, time/memory tradeoff and meet-in-the-middle for a 48-bit key were considered to be impossible one or two decades ago, but can be executed with reasonable effort and cost nowadays, thanks to the rapid advancement of the semiconductor and software industries. In contrast, with enough security margin, many traditional stream ciphers such as HC-256 [40] and Rabbit [9], which were designed roughly at the same time, withstand new attacks and are in use today. The final lesson is that, the lack of public scrutiny makes cryptosystems more vulnerable. Stated simply, the security of a cryptosystem should depend solely on the secrecy of the key, and not on the obscurity of the underlying mechanism.

References

[1] B. Bilgin, A. Bogdanov, M. Knežević, F. Mendel and Q. Wang, Fides: lightweight authenticated cipher with side-channel resistance for constrained hardware, *Cryptographic Hardware and Embedded Systems, CHES'13*, vol. 8086, pp. 142–158, Springer-Verlag, 2013.

[2] J. Borghoff, A. Canteaut, T. Güneysu, E.B. Kavun, M. Knezevic, L. Knudsen, G. Leander, V. Nikov, C. Paar, C. Rechberger and others, PRINCE–a low-latency block cipher for pervasive computing applications, *Advances in Cryptology, ASIACRYPT'12*, vol. 7658, pp. 208–225, Springer-Verlag, 2012.

[3] G. Bard, N. Courtois and C. Jefferson, Efficient methods for conversion and solution of sparse systems of low-degree multivariate polynomials over GF(2) via SAT-solvers, *IACR Cryptology ePrint Archive: Report 2007/024*, pp. 1–14, 2007.

[4] A. Bogdanov, L. Knudsen, G. Leander, C. Paar, A. Poschmann, M. Robshaw, Y. Seurin and C. Vikkelsoe, PRESENT: An ultra-lightweight block cipher, *Cryptographic Hardware and Embedded Systems, CHES'07*, vol. 4727, pp. 450–466, Springer-Verlag, 2007.

[5] A. Biryukov, L. Kizhvatov and B. Zhang, Cryptanalysis of the Atmel Cipher in SecureMemory, CryptoMemory and CryptoRF, *Applied Cryptography and Network Security, ACNS'11*, vol. 6715, pp. 91–109, Springer-Verlag, 2011.

[6] T. Beth and F. Piper, The stop-and-go generator, *Advances in Cryptology, Eurocrypt'84*, vol. 209, pp. 88–92, Springer-Verlag, 1985.

[7] A. Biryukov and A. Shamir, Cryptanalytic time/memory/data tradeoffs for stream ciphers, *Advances in Cryptology, ASIACRYPT'00*, vol. 1976, pp.1–13, Springer-Verlag, 2000.

[8] R. Beaulieu, D. Shors, J. Smith, S. Treatman-Clark, B. Weeks and L. Wingers, The SIMON and SPECK families of lightweight block ciphers, *Cryptology ePrint Archive: Report 2013/404*, pp. 1–45, 2013.

[9] M. Boesgaard, M. Vesterager, T. Pedersen, J. Christiansen and O. Scavenius, Rabbit: A new high-performance stream cipher, *Fast Software Encryption, FSE'03*, pp.307–329, Springer-Verlag, 2003.

[10] N. Courtois, The dark side of security by obscurity and cloning MiFare Classic rail and building passes, anywhere, anytime, *Cryptology ePrint Archive: Report 2009/137*, pp. 1–8, 2009.

[11] N. Courtois and G. Bard, Algebraic cryptanalysis of the data encryption standard, *Cryptography and Coding*, vol. 4887, pp. 152–169, Springer-Verlag, 2007.

[12] J. Balasch, B. Gierlichs, R. Verdult, L. Batina and I. Verbauwhede, Power analysis of Atmel CryptoMemory–recovering keys from secure EEPROMs, *Topics in Cryptology, CT-RSA'12*, vol. 7178, pp. 19–34, Springer-Verlag, 2012.

[13] C. De Canniere, O. Dunkelman, and M. Knežević, KATAN and KTANTAN – a family of small and efficient hardware-oriented block ciphers,

Cryptographic Hardware and Embedded Systems, CHES'09, vol. 5747, pp. 272–288, Springer-Verlag, 2009.

[14] Q, Chai, Design and analysis of security schemes for low-cost RFID systems, *PhD. Thesis, University of Waterloo*, 2012.

[15] N. Courtois and K. Nohl and S. O'Neil, Algebraic Attacks on the Crypto-1 stream cipher in MiFare Classic and oyster cards, *IACR Cryptology ePrint Archive: Report 2008/166*, pp. 1–3, 2008.

[16] N. Courtois, S. O'Neil and J. Quisquater, Practical algebraic attacks on the Hitag-2 stream cipher, *Information Security*, vol. 5735, pp. 167–176, Springer-Verlag, 2009.

[17] M. David, D.C. Ranasinghe, T. Larsen, A2U2: a stream cipher for printed electronics RFID tags, *IEEE International Conference on RFID, RFID'11*, pp. 176–183, 2011.

[18] I. Dinur and A. Shamir, Cube attacks on tweakable black box polynomials, *Advances in Cryptology, Eurocrypt'09*, vol. 5479, pp. 278–299, Springer-Verlag, 2009.

[19] X. Fan, K. Mandal and G. Gong, WG-8: A lightweight stream cipher for resource-constrained smart devices, *The 9th International Conference on Heterogeneous Networking for Quality, Reliability, Security and Robustness, Qshine'13*, vol. 115, pp. 617–632, Springer-Verlag, 2013.

[20] J. D. Golić, Cryptanalytic attacks on MiFare classic protocol, *Topics in Cryptology, CT-RSA'13*, vol. 7779, pp. 239–258, Springer-Verlag, 2013.

[21] S.W. Golomb and G. Gong, *Signal design with good correlation: for wireless communications, cryptography and radar applications*, Cambridge University Press, 2005.

[22] F. D. Garcia, G. Gans, R. Muijrers, P. Rossum, R. Verdult, R. Schreur and B. Jacobs, Dismantling MiFare Classic, *13th European Symposium on Research in Computer Security, ESORICS'08*, vol. 5283, pp. 97–114, Springer-Verlag, 2008.

[23] G. Gans, J. Hoepman and F. D. Garcia, A practical attack on the MiFare Classic, *8th Smart Card Research and Advanced Application Conference, CARDIS'08*, vol. 5189, pp. 267–282, Springer-Verlag, 2008.

[24] J. Guo, T. Peyrin, A. Poschmann and M. Robshaw, The LED block cipher, *Cryptographic Hardware and Embedded Systems, CHES'11*, vol. 6917, pp. 326–341, Springer-Verlag, 2011.

[25] F. D. Garcia, P. Rossum, R. Verdult and R. Schreur, Wirelessly pickpocketing a MiFare Classic card, *In 30th IEEE Symposium on Security and Privacy, S&P'09*, pp. 3–15, 2009.

[26] F. D. Garcia, P. Rossum, R. Verdult and R. Schreur, Dismantling SecureMemory, CryptoMemory and CryptoRF, *Proceedings of the 17th ACM conference on Computer and Communications Security, CCS'10*, pp. 250–259, 2010.

[27] Y. Luo, Q. Chai, G. Gong and X. Lai, WG-7, a lightweight stream cipher with good cryptographic properties, *IEEE Global Telecommunications Conference, GLOBECOM'10*, pp. 1–6, 2010.

[28] MiniSat 2.0. *An open-source SAT solver package*, available at http://www.cs. chalmers.se/Cs/Research/FormalMethods/MiniSat.

[29] K. Nohl, *Cryptanalysis of Crypto-1*, available at http://www.cs.virginia.edu/ kn5f/MiFare.Cryptanalysis.htm, 2008.

[30] K. Nohl, D. Evans, S. Starbug and H. Plötz, Reverse-engineering a cryptographic RFID tag, *Proceedings of the 17th USENIX conference on Security symposium, USENIX'08*, pp. 185–194, 2008.

[31] S. O'Neil and N. Courtois, Reverse-engineered Philips/NXP Hitag-2 cipher, *Rump Session, Fast Software Encryption, FSE'08*, 2008.

[32] Y. Oren, M. Kirschbaum, T. Popp and A. Wool, Algebraic side-channel analysis in the presence of errors, *Cryptographic Hardware and Embedded Systems, CHES'10*, vol. 6225, pp. 428–442, Springer-Verlag, 2010.

[33] H. Plotz and K. Nohl, Breaking Hitag-2, *available at https://har2009.org/program/events/135.en.html, HAR2009*, 2009.

[34] S. Sun, L. Hu, Y. Xie and X. Zeng, Cube cryptanalysis of Hitag-2 stream cipher, *In 10th International Conference on Cryptology and Network Security, CANS'11*, vol. 7092, pp. 15–25, Springer-Verlag, 2011.

[35] K. Shibutani, T. Isobe, H. Hiwatari, A. Mitsuda, T. Akishita and T. Shirai, Piccolo: an ultra-lightweight block cipher, *Cryptographic Hardware and Embedded Systems, CHES'11*, vol. 6917, pp. 342–357, Springer-Verlag, 2011.

[36] P. Stembera and M. Novotny, Breaking Hitag-2 with reconfigurable hardware. *In 14th Euromicro Conference on Digital System Design, DSD'11*, pp. 558–563, 2011.

[37] M. Soos, K. Nohl, and C. Castelluccia, Extending SAT solvers to cryptographic problems, *In 12th International Conference on Theory and Applications of Satisfiability Testing, SAT'09*, vol. 5584, pp. 244–257, Springer-Verlag, 2009.

[38] R. Verdult, F. D. Garcia and J. Balasch, Gone in 360 seconds: Hijacking with Hitag-2, *Proceedings of the 21st USENIX conference on Security symposium, USENIX'12*, pp. 237–252, 2012.

[39] I.C. Wiener. Philips/NXP Hitag-2 PCF7936/46/47/52 stream cipher reference implementation, *available at http://cryptolib.com/ciphers/hitag2/*, 2007.

[40] H. Wu, A new stream cipher HC-256, *Fast Software Encryption, FSE'04*, pp. 226–244, Springer-Verlag, 2004.

In: Radio-Frequency Identification (RFID) ISBN: 978-1-63117-750-7
Editor: Xinxin Fan, pp. 187-213

Chapter 8

GROUPING PROOF PROTOCOLS BASED ON LOW-COST RFID TAGS

***Kuo-Hui Yeh*[1,*], *Nai-Wei Lo*[2,†] *and Siao-Mei Chen*[2,‡]**

[1]Department of Information Management,
National Dong Hwa University, Taiwan
[2]Department of Information Management,
National Taiwan University of Science and Technology, Taiwan

Abstract

RFID technology has been considered by hospitals to provide convenience and comfort for patients and reduce medication and treatment errors in recent years. One of possible RFID applications for hospitals is to attach RFID tags to medicine bags or treatment tools and adopt a well-designed grouping proof protocol to record and verify the corresponding medicine bags or treatment tools are in position. This study proposes new RFID grouping proof protocols for low-cost RFID tags. Our security analysis has indicated that the proposed schemes are secure. Since cheap EPCglobal C1G2 RFID tag can be directly utilized by our protocol, the proposed protocol is well suitable for hospitals to improve patient safety and hospital operation performance with marginal cost on implementation and deployment of such RFID-based verification systems.

*E-mail address: khyeh@mail.ndhu.edu.tw
†E-mail address: nwlo@cs.ntust.edu.tw
‡E-mail address: M10009108@mail.ntust.edu.tw

Keywords: RFID, Grouping Proof, EPC C1G2 Tag, HF Tag

1. Introduction

Owing to the low cost and convenience of identifying an object without physical contact, Radio Frequency Identification (RFID) technology delivers innovative, promising and efficient applications in many domains. An RFID system usually consists of three main components: an RF reader, an RF tag, and a backend server. Figure 1 shows the example of a general RFID system. The communication between the backend server and the reader are wired and could be secured via conventional cryptography mechanisms while the communication between the tags and the reader are wireless and vulnerable to potential malicious attacks. In the normal process of an RFID system, the reader first interrogates the target tag from a tag group. Next, the target tag presents its identification number with other required information to the reader as a response. The reader then communicates with the server for the verification of the information sent from the tag.

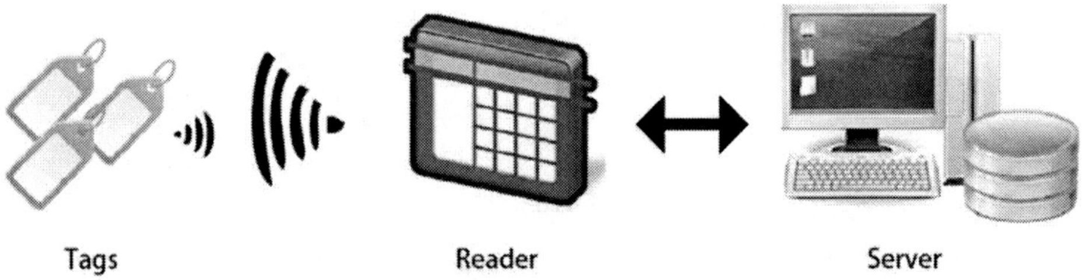

Figure 1. The RFID system model.

According to the investigation from the Institute of Medicine of US National Academies [21], approximate 98,000 people are involved with lethal medication errors every year in US. There are many reasons which could lead to such errors, such as fatigue, carelessness and stress [23, 24]. People are easy to make faults under a tired, stressed or exhausted mode. This applies to both doctors and nurses in extremely busy hospital environments, such as the ratio of patient to doctor/nurse is too high. Therefore, it is important and critical to reduce medication errors for enhancing the quality of the hospital administration.

Recently, RFID technology has been adopted by hospitals to provide more convenience and comfort to patients, and to reduce medication and treatment errors during patient care management [3, 8-11, 16]. Among these applications, one of the most popular RFID-enabled concepts is called RFID grouping proof protocol which allows a trusted verifier to collect information from two or more simultaneously present tags, and create corresponding evidence. This evidence can be utilized to prove the multiple RFID tagged objects were once shown at the same time and the same place. This idea was originally proposed by Juels [17] in 2004, where just two tags (or tagged objects) are considered. The author then provides "RFID yoking protocol" as the name of the proposed protocol. Later, researches [1, 8, 18] extended the RFID yoking protocol to be compatible to multiple tags, and simultaneously enhance the protocol robustness. All of these schemes are categorized to "RFID grouping proof". In the following, we list two typical scenarios of the adoption of RFID grouping proof protocol. Note that more applications of grouping proof scheme can be found in [7].

- In a hospital, tags can be embedded with the medicine containers, and each patient may wear RFID-tagged wristband during the time in hospital. A simple application for enhancing the accuracy of medicine management can thus be implemented. Once the patient would like to pick up their medicine suggested by the doctor, the administrator at the pharmacy can utilize a medicine double-checking procedure with the tags embedded on the medicine containers and the patient's RFID-tagged wristband to further prevent the error of medicine management.
- In another scenario, a patient needs a surgery. In this case, three tags are embedded with the patient, the operating table and the doctor. Before the surgery, the doctor needs to confirm whether "these three tags are actually present" and whether "the connection of these three tags is correct" or not. If it does not hold, the surgery must be stopped. This means that the patient on the operating table may be a wrong person who does not need a surgery. In brief, this application can help a doctor to double check whether the patient is the right one who needs a surgery provided the doctor himself/herself.

In the execution of a grouping-proof protocol, the verifier can be categorized in two different modes: online or offline. In online mode, the network connection between the verifier and the reader is always available, and the verifier can communicate with RFID tags via the reader during each transaction. Although it is more straightforward and easier to design grouping proof protocols with the cooperation of an online verifier, offline verifier based grouping proof schemes are more practical and challenging. It is difficult for the reader to keep stable connections with the verifier all the time; for example, in the mobile environments. So far, some researches have focused on the design of grouping proof protocols suitable to offline mode [3, 8, 25].

In this chapter, we analyze some grouping protocols studied previously and point out that these protocols are vulnerable to several security weaknesses. We then propose a novel RFID grouping proof protocols for low-cost RFID tags, which conform to C1G2 standard, to overcome these vulnerabilities. Due to the characteristics of low-cost and robustness, the proposed protocol is well-suitable for hospital environment to improve patient safety and hospital operation performance. Before illustrating the details of our protocols, we present some assumptions which are generally accepted for the design of a grouping proof protocol [24]:

- RFID readers are potentially untrusted. Only the verifier is trusted.
- RFID readers keep a record of proofs for each session. These records cannot be manipulated by the adversary. In the offline case readers must also store private information regarding interrogation challenges obtained from the verifier.
- The verifier is a trusted entity which shares the secret information with the tags such as cryptographic keys. The RFID reader and the verifier share a secure (private and authenticated) channel. In contrast, the channel between the tags and the reader is public and insecure.
- For the design of grouping proof protocol, the focus is generally set on security issues at the protocol layer and not on physical or link layer issues. Problems such as the coupling design, the power-up and collision arbitration processes or the air-RFID interface are not addressed here.

The remainder of this chapter is organized as follows. An overview of related RFID grouping proof protocols is presented in section 2. Our proposed protocols are introduced in section 3. Section 4 shows the security analysis and performance evaluation. Finally, the conclusion is presented in section 5.

2. Related Work

Compared to common authentication protocols, the goal of grouping (or yoking) proof protocols is to generate a verifiable proof for representing a set of RFID tags were simultaneously scanned at one time. This issue was first introduced by Juels [17] in 2004. Later, the yoking proof protocols are extended to deal with multiple tags, resulting in the grouping proof protocols [18, 19].

Recently, various grouping proof protocols have been proposed against possible attacks. Chien et al. [3] stated that if an online verifier is available, it can directly authenticate and verify proof-involved tags; in that case, an RFID authentication protocol will be more simple or efficient than a grouping-proof pro tocol to operate a protocol instance on each tag. Hence, the authors [3] presented an authentication protocol which improves Huang and Ku's grouping-proof protocol [8] for inpatient medication safety. In addition, an offline grouping-proof protocol is proposed [3]. Unfortunately, Lopez et al. [7] demonstrated that both protocols [3, 8] cannot resist to forgery proof attack and replay attack. In [7], the authors analyzed typical attacks towards grouping proof protocols and presented guidelines for designing a secure grouping proof protocol. A protocol called Kazahaya protocol is thus developed. Later, Ma et al. [12] have the following two observations regarding to Kazah aya scheme.

- The Kazahaya protocol can be used to prove the coexistence of two tags. Therefore, the authors extend it to create a verifiable grouping proof for multiple tags.
- Regarding to the adoption of timestamp in Kazahaya protocol, the authors consider that the verifier cannot know the precise transaction time unless it works in online mode.

Hence, Ma et al. [12] proposed an offline grouping proof protocol with trusted timestamps which help the verifier to obtain the precise transaction time. In the proposed scheme, with the cooperation of the clock tag, a trusted timestamp is included in the grouping proof to show the precise time when it was generated. In addition, this timestamp can be verified by the verifier. However, Lopez et al. [7] protocol is vulnerable to man-in-the-middle attack and Ma et al. [12] is vulnerable to Denal of Proof (DoP) attack [5]. In Lopez et al.'s protocol, once the attacker intercepted authentication value and replayed a malicious value over the insecure radio channel during the execution of a legitimate tag and the reader, the legitimate tag would compute an invalid authentication value involving the malicious value and forward to reader. In the end, the reader would construct an invalid proof that verifier could not verify it. In Ma et al. protocol, if the attacker put few illegitimate t ags among legitimate tags, the generated grouping proof would be invalid and the DoP attack was executed successfully by attacker.

On the other hand, Wickboldt and Piramuthu [9] pointed out that the protocols [3] and [8] need to consider several security issues. According to the results from [9], Yu et al. [10] proposed a lightweight binding proof protocol with low-cost tags. The proposed protocol is based on LAMP [13] with less computational power and implementation cost. Yu et al.'s mechanism utilizes only simple logic gates (e.g. AND, XOR, ADD) to achieve the goal of proving that two tags exist in the field simultaneously. Unfortunately, Wu et al. [11] show that Yu et al.'s protocol [10] is insecure against impersonation attacks with a forged proof. For this reason, Wu et al. [11] demonstrate a lightweight binding proof protocol to overcome the identified weaknesses. Moreover, the proposal does not explore any complicated cryptography function.

Lo and Yeh [5] presented the denial of proof (DoP) attack on the generation of grouping proof for RFID tags. An adversary can easily put few illegitimate tags among legitimate tags to destroy the validity of generated grouping proof. Aforementioned literatures [3, 8, 10, 16] are all vulnerable to DoP attack. To defend against DoP attack, Lo and Yeh [5] proposed three grouping proof protocols including Online Verifier Based Protocol (OVBP), Efficient Online Verifier Based Protocol (EOVBP) and Offline Timestamp Server Based Protocol (OTSBP). In OVBP, each tag will be authenticated by its identity and shared key via an online verifier during the proof generation process. In E-OVBP and OTSBP, the DoP attack will be detected during the proof verification phase and the partial proof constructed by legitimate tags can still be verified. Note that in E-OVBP at least one tag (e.g. Tag1 showed in Fig. 2-2) requires to be authenticated with online verifier.

In a normal process of Figure 2-2, the input for each Tag_i is a combination of the values computed by Tag1. When the process of the proof protocol runs to the Tag_n, it needs Tag_1 to compute the last value and forward this value to the reader. Finally, the reader constructs the grouping proof and sends to online verifier. The form of the whole process resembles the star mode showed in Fig. 2-2. This design can resist DoP attack exactly and prevent the generation of invalid grouping-proof. In contrast, the process of grouping proof protocol forms a circle mode showed in Fig. 2-1. In this process, the input of a Tag_i - except the first message in the protocol - should be a combination of the values computed by its predecessors. When the process runs to the Tag_n, Tag_n computes the last value and sends to the reader, and then the reader construct the grouping proof and sends to the verifier. The form of the whole process looks like the circle showed in Fig 2-1. Many p reviously proposed protocols, such as [3], [8] and [12], are under the circle mode. Nevertheless, this mode does not consider DoP attack, and will easily be compromised by attackers to create an invalid grouping proof. According to the analysis, we adopt the star mode design into our protocols to detect DoP attack effectively.

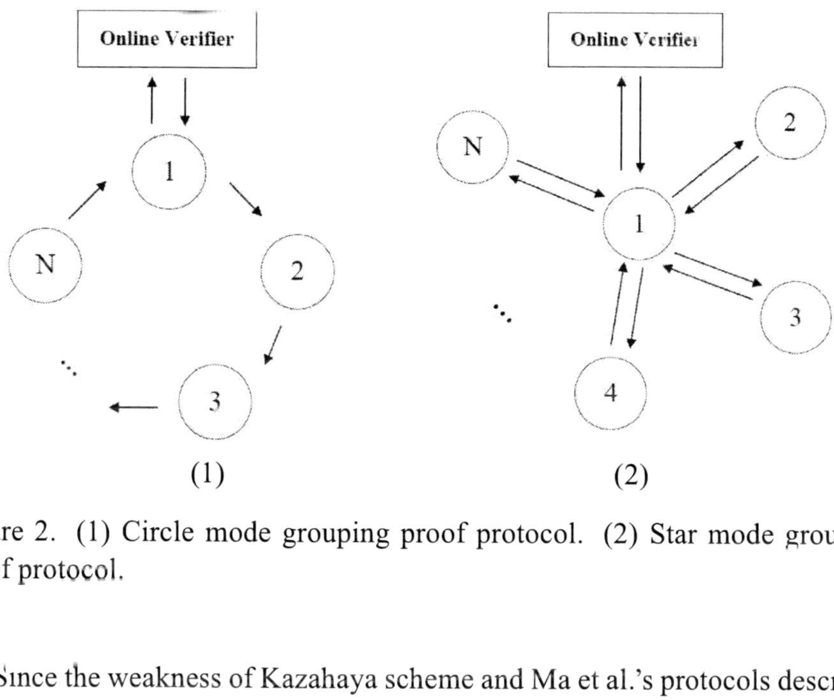

Figure 2. (1) Circle mode grouping proof protocol. (2) Star mode grouping proof protocol.

Since the weakness of Kazahaya scheme and Ma et al.'s protocols described upon, we are motivated to propose a new RFID grouping proof protocol to overcome these vulnerabilities. Our proposed protocol adopts two important designs. (1) the star mode of grouping proof protocol to overcome DoP attack, and (2) the anonymous identifier [10, 11, 13] to prevent the real identifier from being eavesdropped. In addition, we develop a design of privacy protection on proof-involved tags to provide reliable features on tag data security and tag carrier privacy. We show how we use these designs in our proposed protocols and detailed descriptions in next section.

3. The Proposed Protocols

3.1. Description and Assumption

In section 3, we introduce three grouping proof protocols: Online Tag-verification-oriented Protocol (Online TVOP), Online First-tag-authenticated Protocol (Online FTAP), and Offline Tag-verification-oriented Protocol (Offline TVOP).

1. The Online TVOP is designed to work with an online verifier possessing a backend database which stores necessary and related information of all legitimate tags. The crucial feature of Online TVOP is that the verifier (the reader) interacts with tags and authenticate for every proof-involved tag promptly.

2. Similarly, the Online FTAP work with an online verifier which contains a backend database. In contrast to the first protocol, it is based on star mode of grouping proof design, which is described in Chapter 2. The crucial feature of star mode grouping proof is that only one tag needs to be authenticated with the legitimate reader, which is authenticated with verifier at the beginning. This design can prevent the generation of invalid grouping proof and resist to DoP attack. In addition, the Online FTAP produces lower computation cost and get more efficient in executing than the Online TVOP.

3. The Offline TVOP is developed for occasions in which networking (or the verifier and backed servers) is not available or temporarily disconnected. To provide grouping proof under such environment, the Timestamp Generator (TG) components is needed. That is, the TG are assumed to be equipped in each reader when executing the offline TVOP.

Online TVOP and Offline TVOP, tags conform to Gen-2 standard (EPC Class-1 Generation-2, 2008; ISO/IEC 18006-C 2005), which is one of the most relevant standards for low-cost RFID tags. Unlike previous proposals, operations supported on the tags in our protocols are limited to 16-bit pseudorandom number generator (PRNG), bitwise operations (e.g. exclusive OR operation) and cyclic redundancy check (CRC) instead of message authentication code (MAC) and hash functions. In Online FTAP, we adopt HF RFID technology for this protocol. A typical HF RFID system explores passive tags which have a maximum read range of up to 3 feet with faster data rates than LF tags [26, 27]. As the communication range of HF RFID between the reader and a passive tag is short and the data transmission is fast, Online FTAP can resist man-in-the-middle attack. As mentioned in section 1, we list the assumptions adopted for the design of our grouping proof scheme:

- The channel between tags and the reader is considered insecure.
- Adversaries can eavesdrop and modify any message in transmission.
- Once adversaries attempt to spoil a tag to get tag's identifier (or secret key), the tag will be killed by itself and render it silent thereafter.

In our proposals, we assume that each reader has to be authenticated with trusted verifiers (or possesses a trusted TG in advance) before executing the proposed protocols. Since RFID readers are potentially untrusted, the mutual authentication is proposed to ensure that illegitimate readers cannot participate in any proof operation.

3.2. Notations

In this section, the notations are described under the table.

3.3. Mutual Authentication between Verifier/Timestamp Generator and Reader

The steps of the mutual authentication scheme between the verifier and the reader, represented in Fig. 3-1., are:

1. Verifier/TG $\rightarrow$ Reader : "Hello"
2. Reader $\rightarrow$ Verifier/TG : ID_r, a
   The reader computes an encrypted ID_r using reader's secret key K_r.
   $a = E_{K_r}(ID_r)$
   The reader sends ID_r, a to the verifier/TG.
3. Verifier/TG $\rightarrow$ Reader : "Success"
   The verifier/TG get K_r from database according ID_r and computes a^*. If $a^* = a$, the verifier sends "Success" to the reader. Otherwise, the protocol is aborted.

Table 1. The notations throughout this chapter

Notation	Description
Verifier/Database	An online trusted verifier with its own database which contains all legitimate tag information such as unique tag identities and secret key values shared between the tags and the verifier.
Timestamp Generator (TG)	A trusted component with the feature of generating timestamp, which is used in Offline TVOP.
Reader	A legitimate RFID reader.
Tag_i	An RFID tag, which $1 \leqq i \leqq n$ and n is the number of existing tags in a grouping proof session.
t_i /$Timestamp_i$	An encrypted timestamp and a timestamp respectively, where i is a positive number to indicate the sequential order of timestamp generation.
ID_r	A unique identifier of Reader.
ID_i	A unique identifier of Tag_i $(1 \leqq i \leqq n)$.
K_r	The secret key of Reader.
K_i	The secret key of Tag_i $(1 \leqq i \leqq n)$.
K_s	The secret key shared between the verifier and Tag_i $(1 \leqq i \leqq n)$.
r_i / r_i'	Random numbers generated by Tag_i $(1 \leqq i \leqq n)$.
r_s	Random number generated by TG in OFMP.
IDS_i	The anonymity of Tag_i $(1 \leqq i \leqq n)$.
M_i	The tag authentication values computed by Tag_i $(1 \leqq i \leqq n)$.
$M_{1...n}$	The last grouping result computed by Tag_1 in Offline TVOP.
$P_{1...N}$	The grouping proof evidence of $Tag_{1...n}$.
PRNG(s)	A pseudo-random number generator.
$E_k(m)$	Symmetric encryption of message m with secret key K.

The completion of the mutual authentication is builds the trustful connection between the verifier/TG and the reader. In Offline TVOP, the mutual authentication between TG and the reader is also needed to ensure the reader is legitimate.

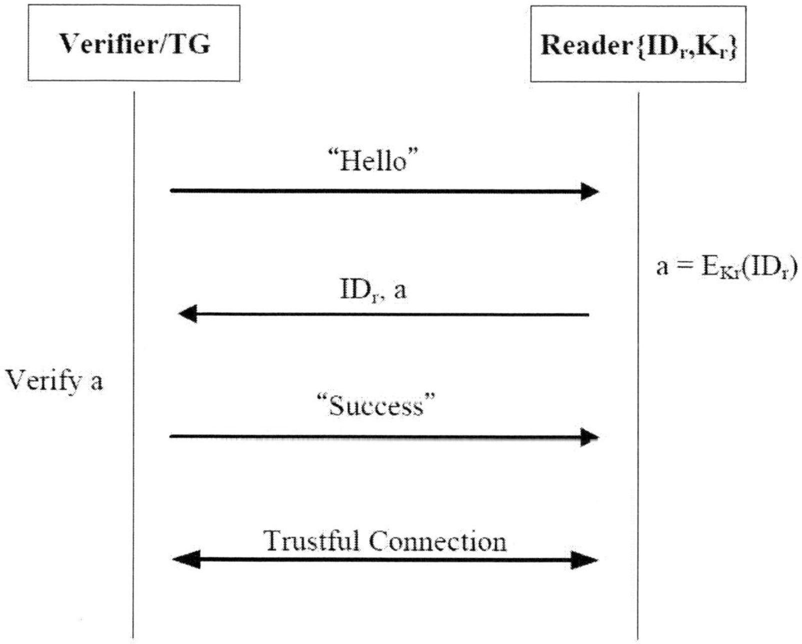

Figure 3.1. The mutual authentication scheme between Verifier/Time Generator and reader.

3.4. The Proposed Grouping Proof Protocols

3.4.1. Online Tag-Verification-Oriented Protocol (Online TVOP)

Online TVOP: Grouping Proof Generation The steps of Online TVOP, represented in Fig. 3-2., are:

1. Verifier $\rightarrow$ Reader : t_1, K_s
   The reader gets an encrypted timestamp t_1=$E_{K_v}(Timestamp)$ from the verifier, where $E_{K_v}(Timestamp)$ denotes an encryption of the current timestamp using verifier's secret key K_v. K_s denotes a secret key shared between the verifier and the tag. Then, the verifier sends $\{t_1, K_k\}$ to the reader.

2. For each Tag_i ($1 \leq i \leq n$), the following steps are executed in order.

   (a) Tag_i $\rightarrow$ Reader : r_i, r'_i, IDS_i, M_i

If i=1, then let M_0=t_1.
Tag_i generates two random numbers r_i, r'_i and computes an anonymity value IDS_i and an authentication value M_i.
IDS_i=$ID_i \oplus PRNG(K_s \oplus r_i)$
M_i=$PRNG(ID_i \oplus r'_i \oplus K_i \oplus M_{i-1}$
Tag_i sends $\{r_i, r'_i, IDS_i, M_i\}$ to the reader.

(b) Reader $\rightarrow Tag_{i+1} : M_i$
After the reader receives values transmitted from Tag_i, the reader verifies Tag_i's authentication value M_i. If the verification is not correct, the protocol is aborted. Otherwise, the reader forwards M_i to the next tag Tag_{i+1}.

3. Reader $\rightarrow$ Verifier : $P_{1...n}$
After the reader visits all tags $\{Tag_1, Tag_2, ..., Tag_n\}$, the reader collects the evidence and constructs a grouping proof $P_{1...n}$. The reader submits the proof $P_{1...n}$ to verifier which can verify it.
$P_{1...n} = \{ID_1, ID_2, ..., ID_n, r'_1, r'_2, ..., r'_n, M_1, M_2, ..., M_n, t\}$

Online TVOP: Grouping Proof Verification The steps of verification phases of the grouping proof in the verifier are:

1. After the grouping proof generated, the verifier gets the current timestamp and estimates the time window using the function described below:
***The estimation of the time window:**
x: The cost of exclusive OR operation.
y: The cost of pseudorandom number generation (PRNG).
According to Online TVOP, a tag's cost is 5x +2y. If the grouping proof contains n tags, the estimated time window is n(5x +2y).
Then, the verifier verifies whether the value of the time difference between the current timestamp and the start timestamp is less than the estimated time window.
$(Thecurrenttimestamp - Thestarttimestamp < Theestimatedtimewindow)$

2. The verifier verifies the validity of grouping proof $P_{1...n} = \{ID_1, ID_2, ..., ID_n, r'_1, r'_2, ..., r'_n, M_1, M_2, ..., M_n, t_1\}$ with the corresponding secret key K_i and identifier ID_i of each Tag_i.

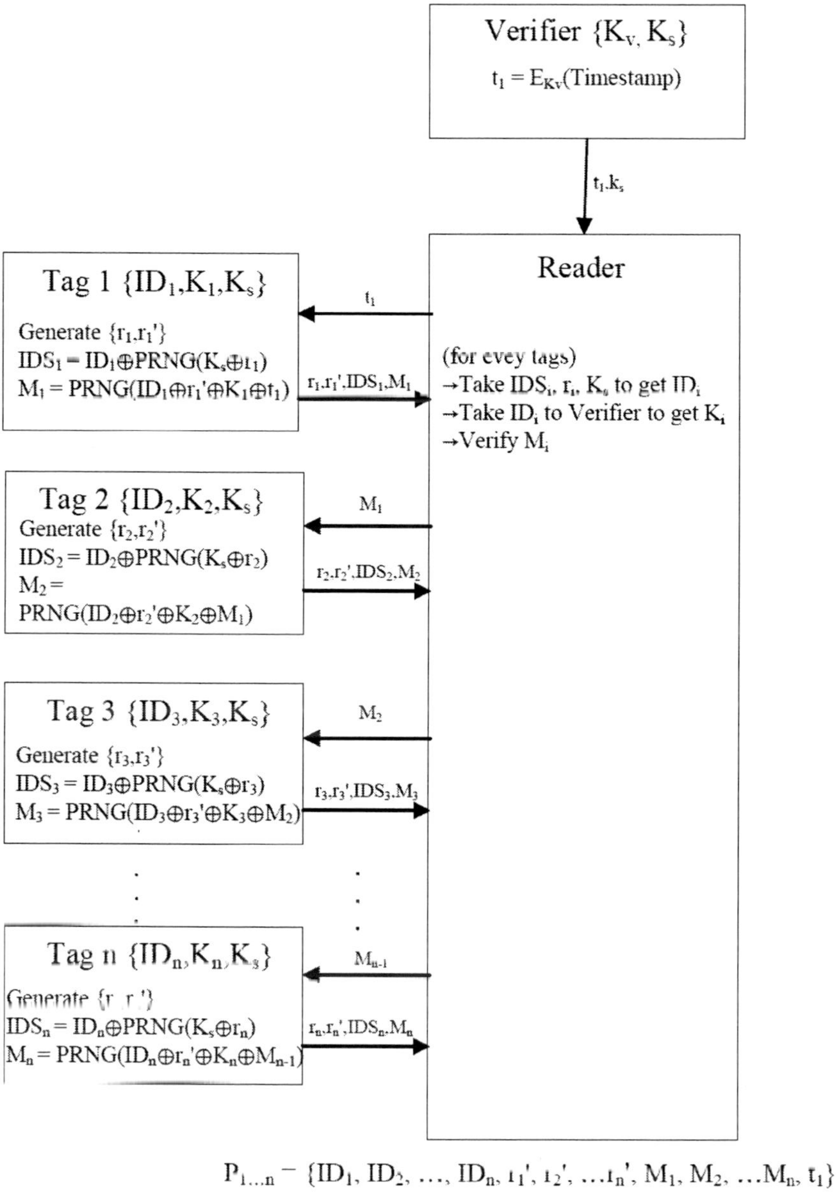

Figure 3.2. Online Tag-verification-oriented Protocol (Online TVOP).

3.4.2. Online First-Tag-Authenticated Protocol (Online FTAP)

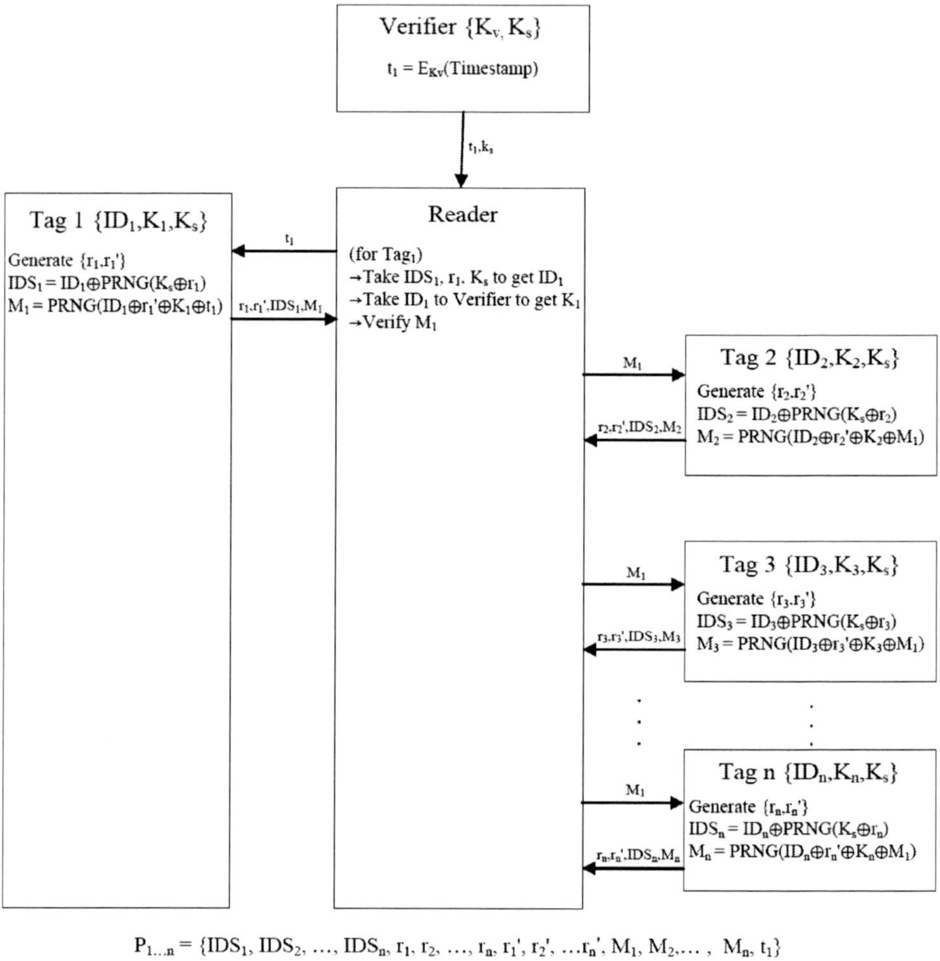

Online FTAP: Grouping Proof Generation The steps of Online FTAP, represented in Fig. 3-3., are:

1. Verifier $\rightarrow$ Reader : t_1, K_s
   First, the reader gets an encrypted timestamp $t_1 = E_{K_v}(Timestamp)$

from the verifier, where $E_{K_v}(Timestamp)$ denotes an encryption of the current timestamp using verifier's secret key K_v. K_s denotes a secret key shared among the verifier and tags. Then, the verifier sends $\{t_1, K_k\}$ to the reader.

2. Reader $\rightarrow Tag_1 : t_1$
   The reader retains K_s and forwards the t_1 to Tag_1.

3. For each Tag_i ($1 \leq i \leq n$), the following steps are executed in order.

   (a) $Tag_i \rightarrow$ Reader : r_i, r'_i, IDS_i, M_i
   If i=1, then let M_0=t_1.
   Tag_i generates two random numbers r_i, r'_i and computes its anonymity value IDS_i and authentication value M_i.
   IDS_i=$ID_i \oplus PRNG(K_s \oplus r_i)$
   M_i=$PRNG(ID_i \oplus r'_i \oplus K_i \oplus M_{i-1}$
   Tag_i sends $\{r_i$, r'_i, IDS_i, $M_i\}$ to the reader.

   (b) Reader $\rightarrow Tag_{i+1} : M_i$
   After the reader receives values transmitted from Tag_i, the reader verifies Tag_i's authentication value M_i when i=1. If the verification is not correct, the protocol is aborted. Otherwise, the reader forwards M_i to the next tag Tag_{i+1}.

4. Reader $\rightarrow$ Verifier : $P_{1...n}$
   After the reader visits all tags $\{Tag_1, Tag_2, ..., Tag_n\}$, the reader collects the evidence and constructs a grouping proof $P_{1...n}$. The reader submits the proof $P_{1...n}$ to verifier which can verify it.
   $P_{1...n} = \{IDS_1, IDS_2, ..., IDS_n, r_1, r_2, ..., r_n, r'_1, r'_2, ..., r'_n, M_1, M_2, ..., M_n, t_1\}$

Online FTAP: Grouping Proof Verification All the steps of verification phases of Online FTAP in the verifier are the same with that in Online TVOP: Once the verifier gets the current timestamp, it estimates the time window via the (1) function x: The cost of exclusive OR operation and (2) y: The cost of pseudorandom number generation (PRNG). Similarly, if the grouping proof scheme contains n tags, the estimated time window is n(5x +2y). Next, the verifier verifies the value of the time difference between the current timestamp and the start timestamp.
$(The current timestamp - The start timestamp < The estimated time window)$

Finally, the verifier examines the validity of grouping proof $P_{1...n} = \{IDS_1, IDS_2, ..., IDS_n, r_1, r_2, ..., r_n, r'_1, r'_2, ..., r'_n, M_1, M_2, ..., M_n, t_1\}$ with the corresponding secret key Ki and identifier ID_i of each Tag_i.

3.4.3. Offline Tag-verification-Oriented Protocol (Offline TVOP)

In this protocol, a trusted TG component installed nearby the reader is used to partially substitute the role of verifier in previous two protocols. The crucial feature of TG is to prevent a legal timestamp from being tampered. All generated grouping proof will be temporarily stored in the reader. When verifier is available, the proofs stored in the reader can be transmitted to and verified by the verifier.

Offline TVOP: Grouping Proof Generation The steps of Offline TVOP, shown in Fig. 3-4., are:

1. Timestamp Generator (TG) $\rightarrow$ Reader : t_1, K_s, a
   First, the TG computes an encrypted timestamp $t_1 = E_{K_v}(Timestamp_1)$, where $E_{K_v}(Timestamp_1)$ denotes an encryption of the current timestamp using shared secret key K_s. TG generates random number r_s and uses r_s to compute the value a., i.e. $a = K_s \oplus r_s$. Then, the verifier sends $\{t_1, K_s, a\}$ to the reader.

2. Reader $\rightarrow Tag_1 : t_1, a$
   The reader retains K_s and forwards the $\{t_1, a\}$ to Tag_1.

3. For each Tag_i ($1 \leq i \leq n$), the following steps are executed in order.

   (a) $Tag_i \rightarrow$ Reader : r_i, r'_i, IDS_i, M_i
   If i=1, then let M_0=t_1.
   Tag_i generates two random numbers r_i, r'_i. Then, Tag_i use $a \oplus K_s$ to gets r_s and computes its anonymity value IDS_i and authentication value M_i. IDS_i=$ID_i \oplus PRNG(r_s \oplus r_i)$
   M_i=$PRNG(ID_i \oplus r'_i \oplus K_s \oplus M_{i-1}$
   Tag_i sends $\{r_i$, r'_i, IDS_i, $M_i\}$ to the reader.

   (b) Reader $\rightarrow Tag_{i+1} : M_i$
   After the reader receives values transmitted from Tag_i, the reader

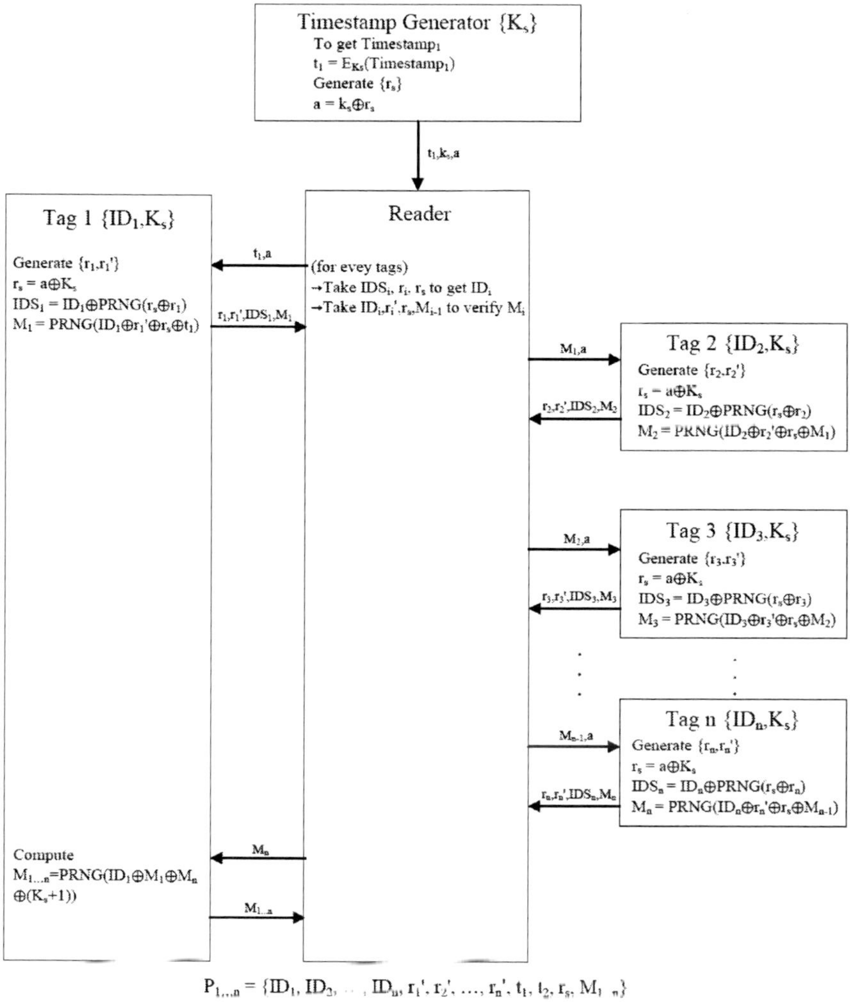

Figure 3.4. Offline Tag-verification-oriented Protocol (Offline TVOP).

verifies Tag_i's authentication value M_i. If the verification is not correct, the protocol is aborted. Otherwise, the reader forwards M_i to the next tag Tag_{i+1}. If $i = n$, the reader forwards M_i to Tag_1 and go to step (4).

4. $Tag_i \rightarrow$ Reader :$M_{1...n}$

Tag_i computes $M_{1...n}$ and sends to the reader. $M_{i...n} = PRNG(ID_i \oplus M_1 \oplus M_n \oplus (K_s + 1\}$

5. Reader $\rightarrow$ Verifier : $P_{1...n}$
After the reader visits all tags $\{Tag_1, Tag_2, ..., Tag_n\}$ and backs to Tag_1, the reader requests TG to obtain the encrypted current timestamp $t2 = E_{K_s}(Timestamp_2)$. Then, the reader collects the evidence and constructs a grouping proof $P_{1...n}$. The reader stores the proof $P_{1...n}$ and submits to verifier later when it becomes available.
$P_{1...n} = \{ID_1, ID_2, ..., ID_n, r'_1, r'_2, ..., r'_n, t_1, t_2, M_{1...n}\}$

Offline TVOP: Grouping Proof Verification The steps of verification phases of the grouping proof in the verifier are:

1. After the grouping proof generated, the verifier decrypts the $\{t_1, t_2\}$ to get $Timestamp_1$ and $Timestamp_2$ and estimates the time window using the functions x and y.
***The estimation of the time window:**
x: The cost of exclusive OR operation.
y: The cost of pseudorandom number generation (PRNG).
According to Offline TVOP, a tag's cost is 6x +2y, and the cost of the last grouping result is 3x + y. If the grouping proof contains n tags, the estimated time window is n(6x +2y) + (3x + y).
Then, the verifier verifies whether the value of the time difference between $Timestamp_1$ and $Timestamp_2$ is less than the estimated time window.
($Timestamp_2 - Timestamp_1 < Theestimatedtimewindow$)

2. The verifier verifies the validity of grouping proof $P_{1...n} = \{ID_1, ID_2, ..., ID_n, r'_1, r'_2, ..., r'_n, t_1, t_2, M_{1...n}\}$ with the corresponding secret key K_i and identifier ID_i of each Tag_i.

4. Security Analysis and Performance Evaluation

In this section, we present the security analysis and performance evaluation of our proposed protocols.

4.1. Security Analysis

We show that the proposed scheme can resist certain possible attacks. We assume that an adversary can intercept and eavesdrop all messages communicated among the tags and the reader.

ID-theft attacks From our assumption, the adversary can eavesdrop any message communicated between the tags and the reader. If the message eavesdropped by adversary is tag's identifier in plaintext, it will be susceptible to traceability attack. For example, in several grouping-proof proposals, tag identifier is sent in plaintext [1, 2, 3]. In another case, Chien and Liu [4] is susceptible to traceability attack due to the involution property of XOR operator. In our three protocols, tag's identifier is sent in ciphertext and independent for different sessions. This design avoids traceability of tag (or tag carrier). We adopt fresh nonces r_i and the secret key, i.e. K_v, of verifier to compute IDS_i to hide the real ID_i. Although that adversary can eavesdrop IDS values, he/she cannot reveal the tag's real ID.

Impersonation tag attack. To perform this attack, an adversary has to eavesdrop the public messages transmitted over the insecure radio channel in current session. Then, the captured messages are replayed to any tag in next session to build a impersonation tag proof. In order to prevent this attack, the computations of a tag - except the first message in the protocol - should be dependent from the values computed by other tags participating in this group. In our three protocols, the input of Tag_i is a combination of the values computed by its predecessors $\{Tag_1, ..., Tag_{i-2}, Tag_{i-1}\}$. Furthermore, we estimate the time window according to the number of proof-involved tags to ensure the time validity of the completion of each generated proof. This can guarantee all proof-involved tags have been scanned simultaneously at the same space.

Forgery proof attack. In forgery proof attack, adversary may utilize the protocol flaws to create a fake proof and spoof the verifier. In our three protocols, we estimate the time window to ensure that each proof is in a limited time window. The computation of the estimated time window is based on the number of tags and the cost of the protocol. In the verification phase of three protocols, the verifier will verify whether the value of $Timestamp_{end} - Timestamp_{start}$ is less than the estimated time window. This mechanism is exactly immune against forgery proof attack.

Replay attack. Our proposed protocols does not suffer from such attacks, because each message transmitted between a reader and a tag is varied with the fresh *nonces* and each session is corresponding to the current timestamp. The adversary might replay the old message but the receiver can detect it because the nonce involved is not the same as the old one and its old authenticator (M_i) is no longer valid now.

Denial of proof (DoP). In [5], the effect of DoP attack is to make generated grouping proof invalid and a DoP attack can be easily executed by putting few illegitimate tags among legitimate tags which are the potential candidates of grouping proof. In Online TVOP and Offline TVOP, we propose authentication mechanism for all proof-involved tag. The reader receives the messages send from a proof-involved tag, and authenticates whether the messages is valid promptly. In Online FTAP, it is based on star mode grouping proof. The crucial feature of star mode grouping proof is that at least one tag needs to be authenticated with the legitimate reader. This design can resist to DoP attack. Based on the above analysis, we believe that our proposals are secure against Denial of Proof attack.

Man-in-the-middle attack. In a normal session of RFID system, the adversary may locate between a legitimate tag and a legitimate reader, and intercept the message transmitted from the tag or the reader. Then, the adversary can replay this message (or modified message) to impersonate a valid entity. The man-in-the-middle attack is launched. In Online TVOP and Offline TVOP, we propose authentication mechanism for all proof-involved tag. The reader receives the message, which is send from a proof-involved tag, and authenticate whether the messages is valid promptly. Therefore, the man-in-the-middle attack cannot be successfully performed. In Online FTAP, we assume that this protocol use the HF tag, which need to close nearly to the reader. According to this feature, the adversary cannot intercept and replay quickly than the transmission. Therefore, this protocol is immune against man-in-the-middle attack.

Table 2. Security comparison between our proposed protocols and related studies

	TA	IA	FA	RA	DoP	MA
[2]Huang and Ku [8]	X	V	X	V	X	X
[2]Chien et al. [3]	X	V	X	V	X	X
[1]Yu et al. [10]	V	V	V	V	X	X
[1]Wu et al. [11]	V	V	V	V	X	X
[1]Peris-Lopez et al. [7]	V	V	V	V	X	X
[2]Ma et al. [12]	V	V	V	V	X	X
[2]Lo et al. OVBP [5]	V	V	V	V	V	V
[2] Lo et al. EOVBP [5]	V	V	V	V	V	V
[2] Lo et al. OTSBP [5]	V	V	V	V	V	V
[2]Proposed Online TVOP	V	V	V	V	V	V
[2]Proposed Online FTAP	V	V	V	V	V	V
[2]Proposed Offline TVOP	V	V	V	V	V	V

TA: ID-theft attack FA: Forgery proof attack MA: Man-in-the-middle attack
IA: Impersonation tag attack RA: Replay attack

[1] Two tag grouping-proof protocol [2] Multitag grouping-proof protocol
V: Prevent successfully. X : Prevent failed.

4.2. Performance Analysis

In this section, we compare our protocols with previous protocols in terms of security comparison and performance comparison. In the Table 2, we compare six known attacks on our protocols and some previous protocols published recently. Our protocols can resist known attacks where other protocols - except Wu et al.'s and Lo et al.'s protocols - may suffer from these attacks. In the Table 3, we show the quantities of operations in our protocols and other related protocols. We summarize all the operations from those grouping proof protocol, and calculate the cost of each operation in fixed data length in the Table 4. The fixed data

Table 3. Performance comparison of grouping proof protocols

	#One way Hash function	#MAC	#CRC	#PRNG	#XOR operation	#Add operation
Huang and Ku [8] [23]	-	-	1	2	2	-
Chien et al. [3] [234]	-	-	-	3	2	-
Yu et al. [10] [1]	-	-	-	-	1	3
Wu et al. [11] [1]	-	-	-	112	78	-
Lopez et al. [7] [1]	-	-	-	9	9	-
Ma et al. [12] [234]	-	-	-	9	10	-
Lo et al. OVBP [5] [23]	1	1	-	-	1	-
Lo et al. EOVBP [5] [23]	1	1	-	1	1	1
Lo et al. OTSBP [5] [234]	-	1	-	1	1	1
Proposed Online TVOP[23]	-	-	-	2	5	-
Proposed Online FTAP[23]	-	-	-	2	5	-
Proposed Offline TVOP[234]	-	-	-	2	6	-
[1] Two tag grouping-proof protocol [2] Multi-tag grouping-proof protocol. [3]We consider that the proof only involves two tags. [4]Offline version.						

Table 4. The cost of each operation in fixed data length

Operation	Cost (ms)
One Way Hashing Function (for MD5)	14.853
One Way Hashing Function (for SHA1)	15.111
Message Authentication Code (MAC) (for MD5)	326.454
Message Authentication Code (MAC) (for SHA1)	312.414
Cyclic redundancy check (CRC)	0.188
Pseudorandom Number Generator (PRNG)	19.417
XOR operation	0.089
Add operation	0.086
The fixed data length is =192 (bits)	

length is refer to the EPC tag specification, which shows the identifier's length is 96 bits [28]. Since many grouping proof protocols did not mention the length of message length they use, we assume the fixed data length (96 bits) instead. In

Table 5. The total cost of each grouping protocol based on each quantities of operations

From	Total Cost (ms)
Huang and Ku [8]	39.201
Chien et al. [3]	58.430
Yu et al. [10]	0.349
Wu et al. [11]	2181.670
Peris-Lopez et al. [7]	175.557
Ma et al. [12]	175.646
Lo et al. OVBP [5]	341.396
Lo et al. EOVBP [5]	360.899
Lo et al. OTSBP [5]	346.046
Lo et al. OVBP [5] for SHA1	341.396
Lo et al. EOVBP [5] for SHA1	347.118
Lo et al. OTSBP [5] for SHA1	332.007
Proposed Online TVOP	39.281
Proposed Online FTAP	39.281
Proposed Offline TVOP	39.370

addition, several grouping proof protocol's operation did not demonstrate what technique is used, we assume the MD5 technique as the one way hashing function and MAC. According to quantities of operations showed in the Table 3, we calculate the total cost of each grouping proof protocol shown in the Table 5 In the Table 5, we present that the performance of our proposed protocols are better than Wu et al.'s and Lo et al.'s protocols even though Wu et al.'s and Lo et al.'s protocols did not suffer from any known attack in our analysis. Based on the above analysis, we can see that our schemes can provide better security guarantee s and require lower computational power and implementation costs.

Conclusion

In this chapter, we have demonstrated that Peris-Lopez et al.'s protocol [7] called kazahaya is vulnerable to man-in-the-middle attack. To overcome these weaknesses, we have proposed new RFID grouping proof protocols for low-cost RFID tags, called Online TVOP, Online FTAP and Offline TVOP, which achieve security enhancement and robust privacy protection. Among the proposed protocols, we adopt the following designs: (1) the tag authentication mechanism with star mode to resist man-in-the-middle attack and DoP attack and (2) the secure mechanisms on tag (or tag carrier) to provide reliable features on tag data security and tag carrier privacy protection. The security and performance analysis has shown that the proposed protocol can provide better security guarantees and can be implemented on low-cost tags. In addition, since the proposed protocols are compatible to EPC Class-1 Generation-2 and Class-1 HF standard, the proposed protocols are well-suitable for hospital environment to improve patient safety and hospital operation performance with reasonable cost on implementation and deployment of RFID-based verification systems.

Acknowledgment

The authors gratefully acknowledge the support from Taiwan Information Security Center (TWISC) and National Science Council, Taiwan, under the Grants Numbers NSC 102-2218-E-259-004 and NSC 102-2218-E-011-013.

References

[1] Piramuthu S., "On Existence Proofs for Multiple RFID Tags," In *Proceedings of the 2006 IEEE International Conference on Pervasive Services*, pp.26-29, June 2006.

[2] Lin C-C, Lai Y-C, Tygar JD, Yang C-K, Chiang C-L, "Coexistence proof using chain of timestamps for multiple RFID tags," In *Proceedings of the 2007 International workshop on database management and application over networks*, pp. 634-643, June 2007.

[3] Chien, H.-Y., Yang, C.-C., Wu, T.-C., Lee, C.-F., "Two RFID-based solutions to enhance inpatient medication safety," *Journal of Medical Systems*, vol.35, pp. 369-375, June 2011.

[4] Chien, H.-Y., Liu, S.-B., "Tree-based RFID yoking proof," In Proceedings of the 2009 International Conference on Networks Security, *Wireless Communications and Trusted Computing*, vol.1, pp.550-553, April 2009.

[5] Lo, M.W., Yeh, K.H., "Anonymous Coexistence Proofs for RFID Tags," *Journal of Information Science and Engineering*, vol. 26, no. 4, pp. 1213-1230, July 2010.

[6] EPCglobal Inc, http://www.gs1.org/gsmp/kc/epcglobal/uhfc1g2/uhfc1g2_1_2_0-standard-20080511.pdf. *Class-1 Generation-2 UHF RFID Protocol for Communications at 860 MHz - 960 MHz* (version 1.2.0).

[7] P. Peris-Lopez, A. Orfila, et al., "Flaws on RFID grouping-proofs. Guidelines for future sound protocols," *Journal of Network and Computer Applications*, vol. 34, no. 3, pp. 833-845, May 2011.

[8] Huang, H.-H., Ku, C.-Y.,"A RFID grouping proof protocol for medication safety of inpatient," *Journal of Medical Systems*, vol. 33, pp. 467-474, December 2009.

[9] Wickboldt A.-K., Piramuthu, S.,"Patient safety through RFID: Vulnerabilities in recently proposed grouping protocols," *Journal of Medical Systems*, vol. 36, pp. 431-435, April 2010.

[10] Yu, Y.-C., Hou T.-W., Chiang, T.-C., "Low Cost RFID Real Lightweight Binding Proof Protocol for Medication Errors and Patient Safety," *Journal of Medical Systems*, vol. 36, pp. 823-828, April 2012.

[11] Wu, S., Chen, K., Zhu, Y.,"A Secure Lightweight RFID Binding Proof Protocol for Medication Errors and Patient Safety," *Journal of Medical Systems*, vol. 36, pp. 2743-2749, October. 2012.

[12] Ma, C., Lin, J., Wang, Y., Shang, M., "Offline RFID Grouping Proofs with Trusted Timestamps," In *Proceedings of the IEEE 11th International Conference on Trust, Security and Privacy in Computing and Communications*, pp. 674-681, June 2012.

[13] Peris-Lopez, P., Hernandez-Castro, J. C., Estevez Tapiador, J. M., and Ribagorda, A., *"LMAP: A real lightweight mutual authentication protocol for low-cost RFID tags," In Proceeding of Workshop on RFID Security - RFIDSec 06*, July 2006.

[14] Barasz, M., Boros, B., Ligeti, P., Loja, K., and Nagy, D., "Breaking LMAP," In *Proceeding of Workshop on RFID Security - RFIDSec* 07, July 2007.

[15] Peris-Lopez, P., Hernandez-Castro, J. C., Estevez Tapiador, J. M., and Ribagorda, A., "LAMED – A PRNG for EPC Class-1 Generation-2 RFID specification," *Journal of Computer Standards and Interfaces*, vol. 31, pp. 88-97, Jan. 2009.

[16] Peris-Lopez, P., Orfila, A., Mitrokotsa, A., and van der Lubbe, J. C. A., "A comprehensive RFID solution to enhance inpatient medication safety," *International Journal of Medical Informatics*, vol. 80, issue 1, pp. 13-24, Jan. 2011.

[17] Juels, A., "Yoking-proofs for RFID tags," In *Proceeding of the 2nd IEEE Annual Conference on Pervasive Computing and Communications Workshops*, pp. 138-143, 2004.

[18] Saito, J., Sakurai, K., "Grouping proof for RFID tags," In *Proceeding of the 19th International Conference on Advanced Information Networking and Applications*, vol. 2, pp. 621-624, 2005.

[19] L. Bolotnyy and G. Robins, "Generalized yoking-proofs for a group of RFID tags," In *Proceeding of the 3rd Annual International Conference on Mobile and Ubiquitous Systems: Networking & Services*, pp. 1-4, 2006.

[20] Kulseng, L., Yu, Z., Wei, Y., Guan, Y., "Lightweight Secure Search Protocols for Low-cost RFID Systems," In *Proceedings of the 2009 29th IEEE International Conference on Distributed Computing Systems*, pp. 40-48, June 2009.

[21] http://www.iom.edu/Reports.aspx.Institute of Medicine of the National Academies.

[22] Crawford, S. Y., Cohen, M. R., Tafesse, E., "Systems factors in the reporting of serious medication errors in hospitals," *Journal of Medical Systems*, vol. 27, Issue 6, pp.543-551, Dec . 2003.

[23] Walton, S. M., "The pharmacist shortage and medication errors: Issues and evidence," *Journal of Medical Systems*, vol.28, pp. 63-59, 2004.

[24] Burmester, M., de Medeiros B, Motta, R., "Provably secure grouping-proofs for RFID Tags," In *Proceedings of the 2008 8th smart card research and advanced applications*, pp. 176-190, September2008.

[25] Leng, X., Lien, Y., Mayes, K., Markantonakis, K., Chiu, J.-H., "Select-response grouping proof for RFID tags," In *Proceedings of the 2009 First Asian Conference on Intelligent Information and Database Systems*, pp.73-77, April 2009.

[26] EPCglobal Inc, http://www.gs1.org/gsmp/kc/epcglobal/uhfc1g2/uhfc1g2_1_2_0-standard-20080511.pdf. *EPC Class-1 HF RFID Air Interface Protocol for Communications at 13.56 MHz* (version2.0.3).

[27] Ching, S.H., Tai, A., "HF RFID versus UHF RFID – Technology for Library Service Transformation at City University of Hong Kong," *The Journal of Academic Librarianship*, vol. 35, pp. 347-359, July 2009.

[28] EPCglobal Inc, http://www.gs1.org/gsmp/kc/epcglobal/tds/tds_1_6-RatifiedStd-20110922.pdf *GS1 EPC Tag Data Standard* (version 1.6)

In: Radio-Frequency Identification (RFID)
Editor: Xinxin Fan, pp. 215-236
ISBN: 978-1-63117-750-7

Chapter 9

SIDE-CHANNEL ATTACKS ON RFID SYSTEMS – STATE OF THE ART

*Xinxin Fan**

Department of Electrical and Computer Engineering
University of Waterloo, Waterloo, ON, Canada

Abstract

Radio Frequency Identification (RFID) systems have become omnipresent and played important roles in both business and private sectors. The increasing reliance on RFID systems for a wide range of applications arises the need to thoroughly investigate the related security mechanisms. Although most modern RFID devices employ cryptographically strong algorithms (e.g., (Triple-)DES or AES) to provide various security services (e.g., encryption, authentication, etc.), the actual hardware realization of RFID devices is subject to side-channel attacks (SCA) that utilize physical properties to break analytically secure cryptographic primitives. In this chapter, we describe a general framework for side-channel attacks on RFID devices and discuss recent progress of applying side-channel attacks to break real-world RFID systems.

Keywords: RFID, security, implementation, measurement tools, side-channel attacks

*E-mail address: x5fan@uwaterloo.ca

1. Introduction

Radio Frequency Identification (RFID) is a ubiquitous wireless communication technology that enables the automatic identification of objects. During the past few years, RFID systems have been widely deployed around the world for numerous large-scale applications such as mobile payment, public transport, access control, e-health, etc., among which many applications are security-relevant due to the processing of sensitive information. A basic RFID system is composed of an RFID reader that is connected to a back-end database and a number of RFID tags. The back-end database stores the concrete information about the tagged objects and is able to be queried using the unique identifier associated with each RFID tag. In an RFID system, the reader usually initiates the communication and the tag equipped with an antenna harvests the power from the RF field of the reader and sends back a response. RFID tags can be categorized into passive, semi-passive, and active tags. While passive tags obtain power directly from the RF field, semi-passive and active tags have an on-board power source. Passive tags can operate in different frequency bands. Low frequency (LF), high frequency (HF), and ultra-high frequency (UHF) tags operate in the 124 – 135 kHz, 13.56 MHz, and 860 – 960 MHz, respectively. With respect to the security functionalities, manufacturers have integrated proprietary cryptographic algorithms (e.g., CRYPTO-1 algorithm [28] and Hitag2 cipher [5]) into LF tags. Unfortunately, the research of the last years have shown that those proprietary algorithms cannot provide sufficient security for RFID systems. For HF tags, strong cryptographic algorithms such as (Triple-)DES or AES have been implemented to provide encryption and authentication. Unlike LF and HF tags, the current generation of UHF tags only provides minimal security through passwords and does not support any cryptographic algorithm.

Although manufacturers have employed mathematically secure cryptographic algorithms for protecting RFID systems, their implementations on RFID tags might be vulnerable to side-channel attacks (SCA). Such attacks measure the side-channel information (e.g., time, power consumption, electromagnetic emanation, etc.) and exploit the fact that the execution of a cryptographic algorithm on an RFID tag leaks information about sensitive information (e.g., secret keys) involved in the computations. When compared to the classical cryptanalysis, side-channel attacks are usually much less general (i.e., target at a specific implementation) but often much more powerful and have emerged as a serious threat for security-oriented RFID systems.

In recent years, a number of side-channel attacks on self-built RFID prototypes as well as commercial RFID tags have been reported in the literature. Those successful attacks demonstrate the vulnerabilities of current secure RFID systems and provide valuable guidance for designing a SCA resistant RFID system in the future. In this chapter, we present a general framework for side-channel attacks on RFID devices and give an overview about recent progress of applying side-channel attacks to break real-world RFID systems. The remainder of this chapter is organized as follows. Section 2 introduces the operation principle of passive RFID systems, followed by the description of different side-channel attacks in Section 3. Section 4 describes the state-of-the-art tools for launching side-channel attacks for RFID systems. In Section 5, we present a general framework for applying correlation power analysis to EM traces collected from RFID tags. Section 6 discusses the recent progress in this field. Finally, Section 7 concludes this chapter.

2. Operation Principle of Passive RFID Systems

In this section, we give a brief description of the most relevant characteristics of the typical passive RFID system operating at 125 kHz, 13.56 MHz, and 860 – 960 MHz, respectively. More comprehensive treatments of the principle and applications of different RFID systems are given in [6] and [11].

2.1. 125 kHz RFID Systems

125 kHz RFID systems operate in the low frequency (LF) band and have been widely used for logistics and animal identification as well as car access and immobilizers. They cover a reading distance from 5 to 100 cm and support a broad range of data rates from 1 KBit/s to 64 KBit/s.

2.1.1. Reader-to-Tag Communication in 125 kHz RFID Systems

Depending on the system design, various modulation schemes such as Amplitude Shift Keying (ASK), Phase Shift Keying (PSK) or Frequency Shift Keying (FSK) can be used to transmit data from the reader to the tag in 125 kHz RFID systems. However, the most commonly used modulation scheme is 100% ASK. Moreover, the reader employs Pulse-Pause Coding (PPC) to provide a relatively continuous energy to the tag. PPC uses fixed gaps to separate the individual data bits, where a logic "1" is considered as a Long and a "0" is a Short burst of RF field between two gaps.

2.1.2. Tag-to-Reader Communication in 125 kHz RFID Systems

Due to the inductive coupling, the energy consumption by the tag has a reactive effect as a voltage drop on the reader. This effect can be used to transfer data from the tag back to the reader by altering a load or a resistance of the integrated circuit in the tag and this process is termed as *load modulation*. For 125 kHz RFID systems, the tag sends back data to the reader by directly modulating the same reader's carrier signal. The two most common codes for tag-to-reader data transmission are the Manchester code and the bi-phase code.

2.2. 13.56 MHz RFID Systems

13.56 MHz RFID systems operate in the high frequency (HF) band and have been widely deployed for access control and contactless smartcard applications. The majority of 13.56 MHz RFID systems (in particular contactless smartcards) on the market is compliant to the ISO/IEC 14443 standard [18], in which a reader is called *Proximity Coupling Device* (PCD) and a contactless smartcard is termed as *Proximity Integrated Circuit Card* (PICC). Typical ISO/IEC 14443 passive tags are designed to operate over a distance of about 10 cm. The ISO/IEC 14443 standard describes the modulation and transmission protocols between PCD and PICC and supports two main communication protocols denoted by Type A and Type B. Both communication schemes support high bit rates of up to 848 kBit/s after the initialization process is completed at a bit rate of 106 Kbit/s.

2.2.1. Reader-to-Tag Communication in 13.56 MHz RFID Systems

The Type A (resp. Type B) modulation scheme uses 100% (resp. 10%) ASK of the RF field for communication from the reader to the tag. The serial bitstream is encoded using the modified Miller code for Type A and the Non Return to Zero (NRZ) code for Type B. Those coding methods can provide a constant level of energy independently from the data transmitted to the tag.

2.2.2. Tag-to-Reader Communication in 13.56 MHz RFID Systems

The data transmission from the tag to the reader is based on the On-Off-Keying (OOK) modulation of a subcarrier with Manchester encoded data. Unlike 125 kHz RFID systems, the baseband data is first encoded using the Manchester code. The Manchester coded data is then modulated to the subcarrier frequency. Finally, the load modulation is applied to the coded subcarrier. For a 13.56 MHz RFID system, a subcarrier of the frequency $f_{sub} = 13.56\ \mathrm{MHz}/16 = 847.5\ \mathrm{kHz}$ is used for the load modulation, which results in the transmitted information being placed in the sidebands of the carrier wave.

2.3. 860 MHz – 960 MHz RFID Systems

860 MHz – 960 MHz RFID systems operate in the ultra-high frequency (UHF) band and are typically utilized to identify products and goods on the retail floor or in the supply chain, where the Electronic Product Code (EPC) that is stored on a UHF RFID tag typically contains product and stocking information. The EPC UHF Class 1 Gen 2 air interface protocol [8], first published by EPCglobal [7] in 2004, defines the physical and logical requirements for a UHF RFID system. The UHF RFID systems support operation at long distances (e.g., 25 – 30 feet) and a tag-to-reader data transfer rate of up to 640 KBit/s.

2.3.1. Reader-to-Tag Communication in 860 MHz – 960 MHz RFID Systems

For the data transmission from the reader to the tag, the digital data is first encoded using the Pulse-Interval Encoding (PIE) scheme. The encoded message is then modulated with one of several variants of the ASK modulation scheme, including double-sideband ASK (DSB-ASK), single-sideband ASK (SSB-ASK), and phase-reversal ASK (PR-ASK) options.

2.3.2. Tag-to-Reader Communication in 860 MHz – 960 MHz RFID Systems

The data transfer from the tag to the reader is based on the *backscatter modulation* of a subcarrier with either FM0 or Miller encoding scheme. In the backscatter modulation, the reader transmits a continuous wave (CW) during the time the reader is listening for the tag signal. The tag sends data back by switching its input impedance between two states and thus modulating the backscattered signal.

Since each impedance state has both a complex and imaginary characteristic, the resulting RF signal shows changes in both amplitude and phase. Therefore, backscattered information from the UHF tag uses a modulation scheme that is a combination of ASK and PSK.

3. Side-Channel Attacks

There exist two main categories of side-channel attacks, namely simple side-channel attacks and differential side-channel attacks. While a simple side-channel attack involves visual inspections of the relation between the side-channel leakage and the secret key stored in the device, a differential side-channel attack is basically a statistical test that examines a large number of data from the side-channel leakage in order to recover a secret key.

3.1. Simple Power Analysis (SPA)

The power consumption of cryptographic devices depends on the processed data and the executed instructions. It can be monitored with an oscilloscope by inserting a small resistor (typically, the resistance value is from 1Ω to 50Ω) in series with the ground or power supply pin. The obtained measurement (i.e., a discrete and quantized time series) is termed as a *power trace*. *Simple power analysis* (SPA) is a side-channel attack which reveals the secret key by observing graphic difference of power consumption traces between different operations and instructions. In a SPA attack, only one or a few power traces are analyzed to recover hidden information. Those information can be the type of an operation, the order of instructions executed, the usage of conditional branches, or in certain cases the secret key. In most cases, a SPA attack requires a precise knowledge about the executed algorithm and its implementation. Moreover, since only a few measurements are performed when realizing a SPA, the quality of the measurements has to be very high. A well-known example of SPA is the square-and-multiply exponentiation algorithm, in which the knowledge whether a multiplication takes place or not directly reveals the currently processed bit of the exponent.

3.2. Differential Power Analysis (DPA)

Differential power analysis (DPA) is an advanced power analysis technique that exploits the correlation between the data and the power consumption of cryptographic algorithms. Since the correlation between data and power consumption is usually attenuated and disturbed by the system noise and cryptographic operations, statistic methods are employed to filter out the noise and magnify the correlation needed for cryptanalysis. In contrast to SPA attacks, DPA attacks do not require the detailed knowledge about the cryptographic device. However, DPA attacks usually need a large number of power traces in order to extract information from extremely noisy traces. Therefore, a precondition of a DPA attack is that an adversary can physically possess and manipulate a cryptographic device for a long time and is able to predict key-dependent intermediate values of the attacked cryptographic algorithm.

The goal of DPA attacks is to recover the secret keys using a large amount of samples of power consumption traces. With the aid of statistic tools DPA attacks can effectively filter out the noise of the attacked cryptographic device and extract the tiny details from power traces that are normally unable to be observed with SPA. In a DPA attack an adversary first feeds a large number of random inputs to the attacked cryptographic device and records the power traces under the unknown secret key. The DPA attack is then taken offline and several key guesses are made. Each guessed key is simulated with the set of inputs used for power measurement to find a proper intermediate value in the algorithm that is strongly related to the power consumption. For each of the key guesses the intermediate value is transformed to the hypothetical power consumption by using different power models. The adversary compares observations of the real power consumption of the attack cryptographic device with the hypothetical power consumption. The correct key is recovered by finding the best match among the measurements and the hypothetical power consumption from different key guesses. Different statistical analysis tools can be employed in DPA attacks and the most commonly used methods include Difference of Means [23], Distance of Means [27] and Correlation Coefficient [3].

3.3. Electromagnetic Analysis (EMA)

Different from power analysis attacks in which an attacker extracts sensitive information through analyzing the power consumption of cryptographic devices, electromagnetic analysis (EMA) attacks exploit the information leaked in the electromagnetic emanations from chips while they are running. Since the flow of electrical power can be viewed as the propagation of electromagnetic waves, the leaked electromagnetic radiation from a chip can directly provide valuable information about the internal operations. EMA attacks are first introduced by Quisquater and Symyde [34] and Gandolfi et al. [12] in 2001 and further developed by Agrawal et al. [1] in 2002. When compared to the power analysis attacks, electromagnetic side-channel is a three dimensional vector field that changes over time. Moreover, the probes used for EMA attacks are different from those for power analysis attacks. Near-field probes are needed in order to capture radiated EM signals and antennas appropriate for particular frequency bands are used to capture far-field signals. The information measured in an EMA attack can also be analyzed in the same way as power consumption, which are termed as *simple electromagnetic analysis* (SEMA) and *differential electromagnetic analysis* (DEMA), respectively.

The EMA attacks, although experimentally more complicated, have several advantages that make EMA attacks more valuable at some point when compared to power analysis attacks. Firstly, EM analysis is a non-invasive attack that measures the EM field emanations. Hence, no direct connection to the cryptographic devices under attack is needed as in the case of power analysis. Secondly, the measurements in the EM analysis usually have a higher signal-to-noise ratio than those collected from the power analysis [12, 34], thereby making the correct key guess easier to identify and requiring fewer samples. Thirdly, the data collected in EM analysis is often better correlated to the processing of secret information than those in power analysis. The reason is that the power analysis involves the power consumption of all components of the chip and in the case of the EM analysis isolating the EM emanations from different components of the chip for individual analysis is possible. Last but not least, it has been shown in [1] that EM emanations consist of a multiplicity of signals, each of which leaks somehow different information about the underlying computations and some information cannot be captured by the power consumption. As a result, some power analysis resistant implementations have been broken by the EM analysis.

4. Tools for Side-Channel Attacks on RFID Systems

For performing side-channel attacks on RFID systems in practice, a range of software and hardware is needed. In this section, we describe the state-of-the-art communication and data acquisition tools that enable an attacker to fully control the communication, recover unknown cryptographic protocols, and automatically collect required data.

4.1. Communication Tools

4.1.1. The Universal Software Radio Peripheral (USRP)

The USRP is a family of software defined radio (SDR) devices designed by Ettus Research [9] for RF applications from DC to 6 GHz. The latest USRP N210 [10] features a Xilinx Spartan 3A-DSP 3400 FPGA, 100 MS/s dual Analog-to-Digital Converter (ADC), 400 MS/s dual Digital-to-Analog Converter (DAC) and Gigabit Ethernet connectivity to transmit data to and from host processors. A modular design allows the USRP to transmit and receive signal in diverse frequency bands with various daughterboards. Moreover, the USRP is programmed and controlled with the open-source software GNURadio [16]. While the USRP and the daughterboard perform the signal up-conversion and down-conversion between the RF frequency to the baseband, the GNURadio implements the modulation, demodulation and other processing of signals. As a general purpose SDR device, the USRP can emulate different RFID readers using the corresponding daughterboards.

4.1.2. The Proxmark III

The Proxmark III [33] is a general all-around RFID testing device for both low (125 kHz – 134 kHz) and high frequency (13.56 MHz) RFID systems. It was originally created by Jonathan Westhues [36] and released under the terms of the general public license. The Proxmark III toolkit consists of the client application running on a PC and two programmable hardware components – a Xilinx Spartan-II FPGA and an ARM7 microcontroller. It has two separate on-board antenna circuits and is able to emulate both an RFID reader and tag, or to eavesdrop the communication between the reader and the tag. The signal from the antenna is routed through the FPGA after it has been digitized by an 8-bit ADC. The FPGA samples the digital signal and passes the sample data to the microcontroller which is responsible for decoding the received samples in order to recover the digital message. For more details about the Proxmark III and its applications for RFID security research, the reader is referred to [14].

4.1.3. Custom RFID Reader and Tag Emulator

A custom, low-cost and freely programmable RFID reader and a versatile tag emulator have been developed by Kasper et al. [19, 20] for emulating the communications in 13.56 MHz RFID systems. The RFID reader, which is built from an ATMega 32 microcontroller, an EM 4094 RF-transceiver, and an ADC, is compliant to ISO/IEC 14443 standard [18]. When compared to the commercially available readers, this device provides full control over the transmitted data down to the bit level, thereby allowing to implement protocols of various contactless smartcards that conform to the ISO standards. The tag emulator, termed as *Chameleon*, is built from an ATxmega192A3 microcontroller and other off-the-shelf discrete components. In particular, the firmware of the Chameleon contains optimized implementations of the Crypto-1 stream cipher, DES, 3DES, and AES, which enables the emulation of the widespread contactless smartcards such as Mifare Classic, Mifare Ultralight, Mifare DESfire, and Mifare DESfire EV1.

4.2. Data Acquisition Tools

4.2.1. Digital Oscilloscope

For acquiring side-channel measurements (e.g., EM emanation, timing, etc.) in RFID systems, a digital oscilloscope that is controlled by a PC is mandatory. A typical oscilloscope used for side-channel attacks on RFID systems (e.g., PicoScope 5000 Series [31]) features 2 or 4 channels, a 60 MHz to 200 MHz input bandwidth, a sampling rate of 1 GHz, an 8-bit to 16-bit vertical resolution, and a 16 MSamples to 512 MSamples waveform memory.

4.2.2. EM Probes and Amplifiers

For measuring EM emanations in RFID systems, near-field probes (e.g., RF2 Near Field Probe Set from Langer EMV-Technik [25]) are usually employed. Those probes differ in measured quantity (i.e., magnetic H-field and B-field), sensitivity, spatial resolution, and frequency bandwidth. B-field probes usually have a coil with multiple windings providing a strong output signal, whereas H-field probes feature an electrically shielded coil with a single winding. For most measurements in RFID systems, H-field probes are usually preferable due to the linear relationship between the current flow in the IC and the output signal of the probe. Furthermore, the EM emanations from passive RFID tags often have small amplitudes in the range of a few millivolts. In order to meet the dynamic range of a typical oscilloscope, an external pre-amplifier (e.g., PA203 or PA303 from Langer EMV-Technik [26]) might be used when necessary. These amplifiers feature almost constant gain of 20 dB or 30 dB over a wide frequency range of 3 GHz.

5. A Framework for Side-Channel Attacks on RFID Systems

In this section, we describe a general framework for performing side-channel attacks on RFID systems, including measurement setup and analysis methodology.

5.1. Measurement Setup

The measurement setup and data acquisition procedure for side channel attacks on RFID systems are illustrated in Figure 1, which contains the following major steps.

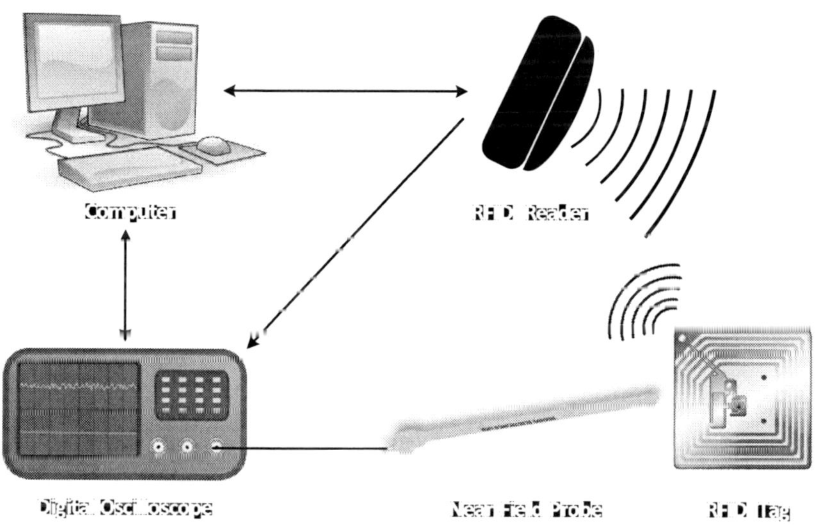

Figure 1. Measurement Setup for Side-Channel Attacks on RFID Systems.

Step 1. The computer sends the configuration data to the oscilloscope through the USB or Ethernet interface. After that, the oscilloscope is armed and waits for the trigger signal to start the measurement.

Step 2. The computer sends a random challenge used in the cryptographic protocol to the RFID reader through the USB or Ethernet interface consecutively. The RFID reader then generates the corresponding security command and switches on the RF field to power up the tag.

Step 3. The RFID reader generates a trigger event to start a measurement with a near-field probe when the last bit of a command has been sent to the tag.

Step 4. The RFID tag conducts the cryptographic operations and transmits its response using the load modulation technique.

Step 5. The RFID reader captures the tag's response and transmits it to the computer. Moreover, the power trace, which is recorded by the oscilloscope, is also sent back to the computer. The power trace together with additional information (e.g., the reader's challenge and the tag's response) will be saved in a file for further analysis.

In the above procedure, Steps 2 – 5 will be performed a number of times (typically a few thousand) until enough power traces have been collected.

5.2. Side-Channel Analysis

For analyzing the susceptibility of an RFID system to side-channel attacks in practice, the most common approach is to apply the correlation power analysis (CPA) to the EM traces collected from an RFID tag, as illustrated in Figure 2.

5.2.1. Random Challenges

Assuming that the required number of power traces is D, a set of D randomly chosen plaintexts/ciphertexts $d_1, d_2, \cdots, d_D$ will be sent to the RFID tag for collecting the EM power traces. Those random inputs will also be fed into the cryptographic algorithm for obtaining the intermediate values. Depending on the specific cryptographic protocols, plaintexts and/or ciphertexts might be needed.

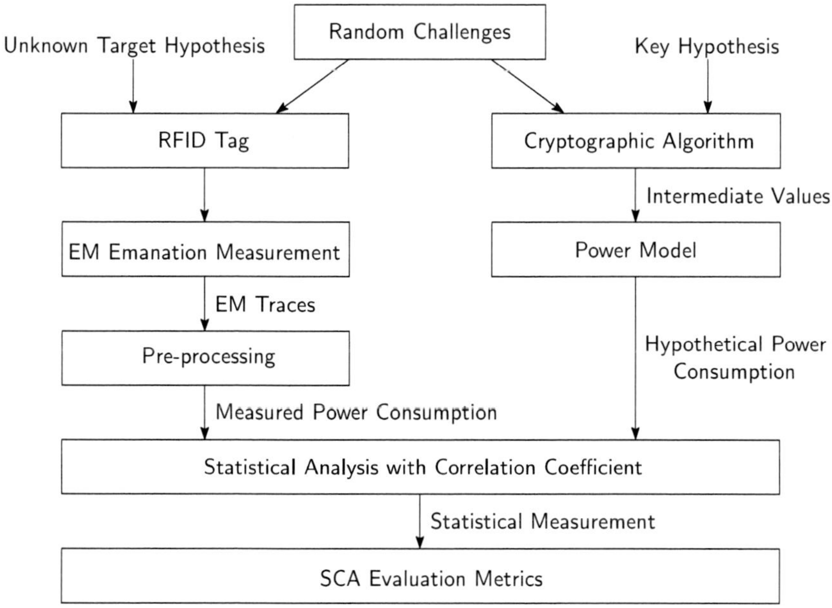

Figure 2. A Framework for Applying CPA Attacks on RFID Systems.

5.2.2. Cryptographic Algorithm and Intermediate Values

Although many manufacturers deployed proprietary cryptographic solutions for securing RFID systems, almost all the secret protocols and algorithms in the current commercial RFID systems have been reverse-engineered during the past few years (see [2, 13, 15, 28] for examples). Therefore, the use of proprietary cryptography and "security by obscurity" should be avoid at any cost when developing secure RFID systems and we should always assume that attackers know all the details of the RFID system except the secret key.

The basic principle of a CPA attack is to hypothesize the influence of one or more bits of the secret key on an intermediate value that depends on the input plaintext/ciphertext d_i and a key hypothesis k_j. Depending on the specific cryptosystems, different intermediate values are eligible for an attack. A non-linear relation between the intermediate values and the key hypothesis (e.g., the output of an S-box in a symmetric-key cipher) is generally beneficial for side-channel analysis. Let $\mathcal{K} = \{k_1, k_2, \cdots, k_K\}$ denote a set of possible partial keys (i.e., k_i is a part of the secret key k). For each plaintext/ciphertext d_i ($i = 1, 2, \cdots, D$) and each key hypothesis $k_j \in \mathcal{K}$ ($j = 1, 2, \cdots, K$), the intermediate value $v_{i,j} = f(d_i, k_j)$ is computed. We represent all intermediate values by a $D \times K$ matrix $\mathcal{V}$:

$$\mathcal{V} = \begin{pmatrix} v_{1,1} & v_{1,2} & \cdots & v_{1,K} \\ \vdots & \vdots & \ddots & \vdots \\ v_{D,1} & v_{D,2} & \cdots & v_{D,K} \end{pmatrix}_{D\times K}.$$

Note that one partial key, denoted by $k^* \in \mathcal{K}$, is the correct key used in the RFID tag. Hence, one column of the matrix $\mathcal{V}$ contains the intermediate values that generated the recorded EM traces at a certain instant in time.

5.2.3. Power Model

A power model, also referred to as a leakage model, is a function that assigns a power consumption predication value to a realization of an intermediate value that is likely to cause information leakage. Depending on the implementation architecture of RFID tags, the most common two power models, namely the Hamming weight (HW) model and the Hamming distance (HD) model, have been widely employed in side-channel analysis and each power model is applicable in certain situation. While the HW model is more suitable for predicting the power consumption of software implementations running on microcontrollers (e.g., contactless smartcards), the HD model is usually used to characterize the power consumption of hardware implementations based on CMOS circuits. As pointed in [27], the extent to which an attacker is able to adequately model power consumption is a major factor for the success of an attack. Using the chosen power model, the transformation of

$$v_{i,j} \rightarrow h_{i,j} \text{ for } i = 1, 2, \cdots, D \text{ and } j = 1, 2, \cdots, K$$

results to the $D \times K$ hypothetical power consumption matrix $\mathcal{H}$:

$$\mathcal{V} = \begin{pmatrix} v_{1,1} & v_{1,2} & \cdots & v_{1,K} \\ \vdots & \vdots & \ddots & \vdots \\ v_{D,1} & v_{D,2} & \cdots & v_{D,K} \end{pmatrix}_{D\times K} \longrightarrow \mathcal{H} = \begin{pmatrix} h_{1,1} & h_{1,2} & \cdots & h_{1,K} \\ \vdots & \vdots & \ddots & \vdots \\ h_{D,1} & h_{D,2} & \cdots & h_{D,K} \end{pmatrix}_{D\times K}.$$

5.2.4. EM Emanation Measurement

Due to the unique communication mechanisms in passive RFID systems, the actual power consumption signal is mixed up with the carrier signal. In order to improve the quality of the exploitable side-channel leakage, Kasper et al. [22] presented the design of an analog demodulator (i.e., a half-wave rectifier) specifically built for refining the side-channel analysis of 13.56 MHz contactless smartcards. The experimental results showed that the designed half-wave rectifier is able to increase the quality of EM measurements and can be easily integrated into the measurement setup.

5.2.5. Pre-processing

Before conducting the statistical analysis, it is important that the collected EM traces are aligned in the time domain. If this is not the case, the number of EM traces required to successfully perform the attack increases significantly. Mangard et al. [27] described the Static Alignment. In this method, the attacker chooses a fragment in a reference EM trace which is close to the position of the leakage and shift other traces so that the reference fragments are aligned. In [37], Woudenberg et al. proposed the Elastic Alignment, which aims to stretch and shrink the trace in such a way that the time dimension is fully restored. However, the matching procedure in this method is quite computationally expensive. Clavier et al. [4] presented the Sliding Window DPA (SW-DPA) that is specifically designed to thwart the side-channel attack countermeasure called random process interrupts (RPI). In [35], Muijrers proposed Rapid Alignment Method (RAM) which is based on the SIFT and U-SURF algorithms for object recognition in images. RAM consists of four major components: 1) A detector for detecting point of interest; 2) A description for generating a feature vector; 3) A matcher for matching points between power traces; and 4) A warper for aligning the trace. In particular, RAM outperforms Static Alignment and SW-DPA in terms of the alignment capability and Elastic Alignment with respect to the alignment speed.

5.2.6. Statistic Analysis with Correlation Coefficient

The matrix $\mathcal{H}$ generated in Section 5.2.3 consists of the hypothetical power consumption values for all D inputs and K possible keys. In order to find the correct key $k^* \in \mathcal{K}$, we compare the two matrices $\mathcal{T}$ and $\mathcal{H}$ column by column by applying the Pearson correlation coefficient. More specifically, we compute the correlation coefficient $r_{j,l}$ of the jth column of $\mathcal{H}$ and the lth column of $\mathcal{T}$

for all $j = 1, 2, \cdots, K$ and $l = 1, 2, \cdots, T$ as follows:

$$r_{j,l} = \frac{\sum_{i=1}^{D}(h_{i,j} - \bar{h}_j)(t_{i,l} - \bar{t}_l)}{\sqrt{\sum_{i=1}^{D}(h_{i,j} - \bar{h}_j)^2 \cdot \sum_{i=1}^{D}(t_{i,l} - \bar{t}_l)^2}},$$

where $\bar{h}_j = \frac{1}{D}\sum_{i=1}^{D} h_{i,j}$ and $\bar{t}_l = \frac{1}{D}\sum_{i=1}^{D} t_{i,l}$. All correlation coefficients are then stored in a $K \times T$ matrix $\mathcal{R}$:

$$\mathcal{R} = \begin{pmatrix} r_{1,1} & r_{1,2} & \cdots & r_{1,T} \\ \vdots & \vdots & \ddots & \vdots \\ r_{K,1} & r_{K,2} & \cdots & r_{K,T} \end{pmatrix}_{K \times T}.$$

The value with the highest correlation coefficient in the matrix $\mathcal{R}$, denoted by $r_{k^*,\tau}$, indicates the correct secret key k^* as well as the time instant of the leakage τ.

6. Side-Channel Attacks on RFID Systems in Practice

In this section, we give a brief overview of recent progress in side-channel attacks on RFID systems in practice.

6.1. Side-Channel Attacks on 13.56 MHz RFID Systems

In [17], Hutter et al. conducted power and EM attacks on two RFID tag prototypes, one of which implements AES on a standard 8-bit microcontroller and the other of which features an AES co-processor. The authors showed that both prototype devices can be successfully attacked with less than 1000 EM traces. However, the antenna and analog front-end are separated from the digital circuit on their prototype devices, which is quite different from real RFID tags. While this work is the first one considering DPA and DEMA attacks on RFID devices, the feasibility of applying those attacks to the real-world situation is still not clear. Since 2009, the research community has started investigating application of side-channel attacks to break commercial and real-world RFID systems.

Mifare DESFire, which was introduced in 2002 by NXP and deployed in several payment and transportation systems around the world, is a contactless smartcard featuring a cryptographic engine for authentication and encryption based on (Triple-)DES. In [21], Kasper et al. conducted the first non-invasive, black-box analysis of the Mifare DESFire contactless smartcard. The authors proposed the leakage model and described a CPA on the Triple-DES hardware implementation on the target smartcard. However, the authors cannot recover the full key at that time and did not disclose the specific device under test. This initial work has been improved significantly in [30], where Osward and Paar detailed the procedure for recovering the complete 112-bit secret key of the Triple-DES algorithm using the CPA and the template attack on the popular Mifare DESFire MF3ICD40 smartcard.

Near Field Communication (NFC) is a short-range, contactless communication technology that operates at 13.56 MHz and has become increasingly popular and enabled numerous applications (e.g., mobile payment, ticketing, transportation, etc.) in the last years. Recently, Korak and Plos [24] demonstrated the feasibility of launching a remote side-channel attack at a distances up to 1 m on a security-enabled NFC tag with an integrated AES module. The experiment setup is similar to that in Figure 1 (see Section 5.1) except that the near-field probe is replaced by a self-made loop antenna. Moreover, the authors characterized the relationship between attack performance and measurement distance and verified it through extensive experiments. This work highlighted the importance of integrating suitable side-channel attack countermeasures for security-enabled NFC devices.

6.2. Side-Channel Attacks on 860 MHz – 960 MHz RFID Systems

In 2007, Oren and Shamir [29] published the first paper about password extraction from EPC Class 1 Generation 1 RFID tags using the remote power analysis attacks. The basic idea for recovering the password is to detect the small fluctuations in the reader field during the communication with the tag, which allows an attacker to consecutively retrieve password bits. Due to the high volume and tight cost, EPC Class 1 Generation 1 RFID tags only have minimal support for security (i.e., static passwords to access or kill information on the tag) and do not implement any cryptographic algorithms.

Later on, Plos [32] investigated the susceptibility of UHF RFID tags using differential EM analysis attacks. With a simple loop dipole antenna and a digital oscilloscope, the authors demonstrated the data-dependent emanation of the UHF RFID tags at distances up to 1 m. Beside the commercially available UHF RFID tags, the authors also built a semi-passive UHF tag prototype based on an 8-bit microcontroller ATMega128, which has been used as a reference for providing the trigger signal when performing measurements on commercial UHF RFID tags. Although the current generation of UHF RFID tags does not use cryptographic protection, this work emphasized that the potential side-channel attacks should be considered in the future design of secure UHF RFID tags.

Conclusion

In this chapter, we provide a comprehensive research survey about side-channel attacks on RFID systems, which include the basic principle of RFID systems and side channel attacks, a general attack framework, the state-of-the-art measurement and data acquisition tools, and the recent research progress. It is our goal to highlight the threat of side-channel attacks on security-oriented RFID applications and to expect that manufacturers and system integrators will be aware of the potential security risks and implement suitable SCA countermeasures in their future design of RFID systems.

References

[1] D. Agrawal, B. Archambeault, J. R. Rao, and P. Rohatgi, "The EM Side-Channel(s)", *The 4th International Workshop on Cryptographic Hardware and Embedded Systems - CHES 2002*, ser. LNCS 2523, B. S. Kaliski Jr., Ç. K. Koç, and C. Paar (eds.), Berlin, Germany: Springer-Verlag, pp. 29-45, 2002.

[2] S. C. Bono, M. Green, A. Stubblefield, A. Juels, A. D. Rubin, and M. Szydlo, "Security Analysis of a Cryptographically-Enabled RFID Device", *The 14th conference on USENIX Security Symposium*, USENIX Association, 2005.

[3] E. Brier, C. Clavier, and F. Olivier, "Correlation Power Analysis with a Leakage Model", *The 6th International Workshop on Cryptographic Hardware and Embedded Systems - CHES 2004*, ser. LNCS 3156, M. Joye and J.-J. Quisquater (eds.), Berlin, Germany: Springer-Verlag, pp. 16-29, 2004.

[4] C. Clavier, J.-S. Coron, and N. Dabbous, "Differential Power Analysis in the Presence of Hardware Countermeasures", *The 2nd International Workshop on Cryptographic Hardware and Embedded Systems - CHES 2000*,

ser. LNCS 1965, Ç. K. Koç and C. Paar (eds.), Berlin, Germany: Springer-Verlag, pp. 252-263, 2000.

[5] N. Courtois, S. O'Neil, and J.-J. Quisquater, "Practical Algebraic Attacks on the Hitag2 Stream Cipher", *The 12th International Conference on Information Security - ISC 2009*, ser. LNCS 5735, P. Samarati et al. (eds.), Berlin, Germany: Springer-Verlag, pp. 167-176, 2009.

[6] D. M. Dobkin. *The RF in RFID: UHF RFID in Practice*, Elsevier Inc., Second edition, 2013.

[7] *EPCglobal.* http://www.gs1.org/epcglobal.

[8] EPCglobal. EPC™*Radio-Frequency Identitiy Protocols Generation-2 UHF RFID – Specification for RFID Air Interface Protocol for Communications at 860 MHz - 960 MHz* (Version 1.0.9), http://www.gs1.org/gsmp/kc/epcglobal/uhfc1g2/uhfc1g2_1_0_9-standard-20050126.pdf, January 2005.

[9] *Ettus Research.* http://www.ettus.com/.

[10] *Ettus Research.* USRP N200/N210 Networked Series. https://www.ettus.com/content/files/07495_Ettus_N200-210_DS_Flyer_HR_1.pdf.

[11] K. Finkenzeller. *RFID Handbook: Fundamentals and Applications in Contactless Smart Cards and Identification*, John Wiley and Sons, 3rd edition, 2010.

[12] K. Gandolfi, C. Mourtel, and F. Olivier, "Electromagnetic Analysis: Concrete Results", *The 3rd International Workshop on Cryptographic Hardware and Embedded Systems - CHES 2001*, ser. LNCS 2162, Ç. K. Koç, D. Naccache, and C. Paar (eds.), Berlin, Germany: Springer-Verlag, pp. 251-261, 2001.

[13] F. D. Garcia, G. K. Gans, R. Muijrers, P. Rossum, R. Verdult, R. W. Schreur, and B. Jacobs, "Dismantling MIFARE Classic", *The 13th European Symposium on Research in Computer Security - ESORICS 2008*, ser. LNCS 5283, S. Jajodia and J. López (eds.), Berlin, Germany: Springer-Verlag, pp. 97-114, 2008.

[14] F. D. Garcia, G. K. Gans, and R. Verdult, "Tutorial: Proxmark, the Swiss Army Knife for RFID Security Research", Tutorial at 8th Workshop on RFID Security and Privacy (RFIDSec 2012), *Technical Report*, Radboud University Nijmegen, 2012.

[15] F. D. Garcia, G. K. Gans, Roel Verdult, and M. Meriac, "Dismantling iClass and iClass Elite", *The 13th European Symposium on Research in Computer Security - ESORICS 2008*, ser. LNCS 7459, S. Foresti, M. Yung, and F. Martinelli (eds.), Berlin, Germany: Springer-Verlag, pp. 697C715, 2012.

[16] *GNURadio*. http://gnuradio.org/redmine/.

[17] M. Hutter, S. Mangard, and M. Feldhofer, "Power and EM Attacks on Passive 13.56 MHz RFID Devices", *The 9th International Workshop on Cryptographic Hardware and Embedded Systems - CHES 2007*, ser. LNCS 4727, P. Paillier and I. Verbauwhede (eds.), Berlin, Germany: Springer-Verlag, pp. 320-333, 2007.

[18] International Organization for Standardization (ISO). *ISO/IEC 14443: Identification Cards - Contactless Integrated Circuit(s) Cards - Proximity Cards* - Part 1-4, 2008.

[19] T. Kasper, D. Carluccio, and C. Paar, "An Embedded System for Practical Security Analysis of Contactless Smartcards", *The First IFIP TC6/WG 8.8/WG 11.2 International Workshop on Information Security Theory and Practices - WISTP 2007*, ser. LNCS 4462, D. Sauveron et al. (eds.), Berlin, Germany: Springer-Verlag, pp. 150-160, 2007.

[20] T. Kasper, I. von Maurich, D. Oswald, and C. Paar, "Chameleon: A Versatile Emulator for Contactless Smartcards", *The 13th International Conference on Information Security and Cryptology - ICISC 2010*, ser. LNCS 6829, K. H. Rhee and D. Nyang (eds.), Berlin, Germany: Springer-Verlag, pp. 189-206, 2010.

[21] T. Kasper, D. Oswald, and C. Paar, "EM Side-Channel Attacks on Commercial Contactless Smartcards using Low-Cost Equipment", *The 10th International Workshop on Information Security Applications - WISA 2009*, ser. LNCS 5932, H. Y. Youm and M. Yung (eds.), Berlin, Germany: Springer-Verlag, pp. 79-93, 2009.

[22] T. Kasper, D. Oswald, and C. Paar, "Side-Channel Analysis of Cryptographic RFIDs with Analog Demodulation", *The 7th International Workshop on RFID Security and Privacy - RFIDSec'11*, ser. LNCS 7055, A. Juels and C. Paar (eds.), Berlin, Germany: Springer-Verlag, pp. 61-77, 2012.

[23] P. C. Kocher, J. Jaffe, and B. Jun, "Differential Power Analysis", *Advances in Cryptology - CRYPTO 1999*, ser. LNCS 1666, M. Wiener (ed.), Berlin, Germany: Springer-Verlag, pp. 388-397, 1999.

[24] T. Korak and T. Plos, "Applying Remote Side-Channel Analysis Attacks on a Security-Enabled NFC Tag", *Topics in Cryptology - CT-RSA 2013*, ser. LNCS 7779, E. Dawson (ed.), Berlin, Germany: Springer-Verlag, pp. 207-222, 2013.

[25] Langer EMV-Technik. *RF2 Near Field Probe Set*, http://www.langer-emv.de/en/products/disturbance-emission/near-field-probes/rf-2/.

[26] Langer EMV-Technik. *Preamplifier PA203 and PA303*, http://www.langer-emv.de/en/products/disturbance-emission/preamplifier/pa-203-303/.

[27] S. Mangard, E. Oswald, and T. Popp, *Power Analysis Attacks: Revealing the Secrets of Smart Cards*, Springer-Verlag, Secaucus, HJ, USA, 2007.

[28] K. Nohl, D. Evans, S. Starbug, and H. Plötz, "Reverse-Engineering A Cryptographic RFID Tag", *The 17th conference on USENIX Security Symposium*, USENIX Association, 2008.

[29] Y. Oren and A. Shamir, "Remote Password Extraction from RFID Tags", *IEEE Transactions on Computers*, Vol. 56, Iss. 9, pp. 1292-1296, IEEE Computer Society, 2007.

[30] D. Oswald and C. Paar, "Breaking Mifare DESFire MF3ICD40: Power Analysis and Templates in the Real World", *The 13th International Workshop on Cryptographic Hardware and Embedded Systems - CHES 2011*, ser. LNCS 6917, B. Preneel and T. Takagi (eds.), Berlin, Germany: Springer-Verlag, pp. 207-222, 2011.

[31] Pico Technology. *PicoScope 5000 Series – Flexible Resolution Oscilloscopes*, http://www.picotech.com/picoscope5000.html.

[32] T. Plos, "Susceptibility of UHF RFID Tags to Electromagnetic Analysis", *Topics in Cryptology - CT-RSA 2008*, ser. LNCS 4964, T. Malkin (ed.), Berlin, Germany: Springer-Verlag, pp. 288-300, 2008.

[33] *Proxmark III*. http://proxmark3.com/.

[34] J.-J. Quisquater and D. Samyde, "Electromagnetic Analysis (EMA): Measures and Counter-Measures for Smart Cards", *International Conference on Research in Smart Cards - E-smart 2001*, ser. LNCS 2140, I. Attali and T. P. Jensen (eds.), Berlin, Germany: Springer-Verlag, pp. 200-210, 2001.

[35] R. Muijrers, J. van Woudenberg, and L. Batina, "RAM: Rapid Alignment Method", *The 10th IFIP WG 8.8/11.2 International Conference on Smart Card Research and Advanced Applications - CARDIS 2011*, ser. LNCS 7079, E. Prouff (ed.), Berlin, Germany: Springer-Verlag, pp. 266-282, 2011.

[36] J. Westhues, *Proxmark III: A Test Instrument for HF/LF RFID*. http://cq.cx/proxmark3.pl.

[37] J. van Woudenberg, M. Witteman, and B. Bakker, "Improving Differential Power Analysis by Elastic Alignment", *Topics in Cryptology - CT-RSA 2011*, ser. LNCS 6558, A. Kiayias (ed.), Berlin, Germany: Springer-Verlag, pp. 104-119, 2011.

INDEX

#

A

B

C

D

E